Engineering Materials

This series provides topical information on innovative, structural and functional materials and composites with applications in optical, electrical, mechanical, civil, aeronautical, medical, bio- and nano-engineering. The individual volumes are complete, comprehensive monographs covering the structure, properties, manufacturing process and applications of these materials. This multidisciplinary series is devoted to professionals, students and all those interested in the latest developments in the Materials Science field, that look for a carefully selected collection of high quality review articles on their respective field of expertise.

Indexed at Compendex (2021) and Scopus (2022)

Syeed SaifulAzry Osman Al Edrus ·
Seng Hua Lee · Juliana Abdul Halip ·
Norfaryanti Kamaruddin ·
Nur Izzah Nabilah Haris · Paridah Md Tahir
Editors

Green Lignocellulosic-Based Panels

Manufacturing, Characterization and Applications

Springer

Editors
Syeed SaifulAzry Osman Al Edrus
Institute of Ecosystem Science Borneo
Universiti Putra Malaysia (Bintulu Campus)
Bintulu, Sarawak, Malaysia

Seng Hua Lee
Department of Wood Industry
Universiti Teknologi MARA Pahang
Branch Jengka Campus
Pahang, Malaysia

Juliana Abdul Halip
Faculty of Engineering Technology
Universiti Tun Hussein Onn Malaysia
(UTHM)
Johor, Malaysia

Norfaryanti Kamaruddin
Institute of Tropical Forestry and Forest
Products (INTROP)
Universiti Putra Malaysia
Selangor, Malaysia

Nur Izzah Nabilah Haris
Department of Chemical
and Environmental Engineering
Faculty of Engineering
Universiti Putra Malaysia
Serdang, Selangor, Malaysia

Paridah Md Tahir
Institute of Tropical Forestry and Forest
Products (INTROP)
Universiti Putra Malaysia
Serdang, Selangor, Malaysia

ISSN 1612-1317 ISSN 1868-1212 (electronic)
Engineering Materials
ISBN 978-981-96-3406-4 ISBN 978-981-96-3404-0 (eBook)
https://doi.org/10.1007/978-981-96-3404-0

This Springer imprint is published by the registered company Springer Nature Singapore Pte Ltd.
The registered company address is: 152 Beach Road, #21-01/04 Gateway East, Singapore 189721, Singapore

Preface

Lignocellulosic components, comprising lignin, cellulose, and hemicelluloses, serve as a crucial renewable resource for the wood-based panels sector. Sourcing from agricultural crop residues, forest residues, and grass materials, they represent the most plentiful biomass resource on Earth and are cost-effective. As the availability of wood with high quality continues to decline globally, the use of lignocellulosic materials is particularly crucial, notably for the manufacture of wood-based panels. The increasing environmental concerns confronting our society today—such as deforestation, resource depletion, and climate change—necessitate a transition to more sustainable materials and practices. This has stimulated an increasing interest in eco-friendly materials sourced from renewable resources. This book, Green Lignocellulosic-Based Panels: Manufacturing, Characterization, and Applications, provides an in-depth examination of the production, characteristics, and uses of sustainable, lignocellulosic panel products. The book aims to be a significant resource for researchers, industry professionals, and students focused on the principles and methods of sustainable materials science. It investigates the capacity of lignocellulosic materials—sourced from wood, agricultural byproducts, and other biomass waste—to diminish reliance on non-renewable resources and mitigate environmental effects.

Lignocellulosic materials can originate from both wood and non-wood sources. Wood encompasses both softwood and hardwood, whereas non-wood comprises a diverse array of biomass apart from wood, such as oil palm, bamboo, and sugarcane. Therefore, the book begins with an introduction to wood and non-wood-engineered products. This foundation provides readers with the necessary understanding of the existing types of engineered products from both wood and non-wood resources. The adhesives employed to bond the composites are crucial for the production of green lignocellulosic-based panels. Adhesives are crucial components in composite manufacturing, significantly influencing both the performance of the composite and its environmental friendliness. In the following chapters, lignin-based adhesive is emphasized. Lignin is the most prevalent aromatic natural compound and has emerged as a possible alternative to petroleum-based polymers, facilitating the development of new materials for wood adhesives. This chapter presents a contemporary overview of the lignin alterations documented in recent years for the synthesis of lignin-based wood adhesives. These improvements address the issues encountered in the use of lignin under various treatments and environments to develop adaptable lignin-based products.

The subsequent chapters elucidate and examine several wood-based goods. Examples include particleboard, fiberboard, veneer-based composites, wood polymer composites (WPC), and sound and thermal insulation panels. Additionally, many categories of green binders were emphasized, including tannin, protein, and glucose. Certain chapters explore the advancement of binderless composites in further detail. Binderless composites have garnered considerable interest among researchers as sustainable and eco-friendly alternatives to conventional composites. Binderless composites eschew synthetic binders, employing natural wood components such as fibers, particles, and veneers that possess intrinsic self-bonding characteristics. Last but not least, the last chapter embarks on the multifaceted nature of the environmental impact of green lignocellulosic-based panels. The economic effects of the green panels were also emphasized. Economic viewpoints in green panels emphasize sustainability, innovation, and resilience. Investigating sustainable materials corresponds with ecological objectives and generates commercial prospects. Effective raw material acquisition and sustainable production methods enhance cost efficiency and competitiveness.

Sarawak, Malaysia — Syeed SaifulAzry Osman Al Edrus
Pahang, Malaysia — Seng Hua Lee
Johor, Malaysia — Juliana Abdul Halip
Selangor, Malaysia — Norfaryanti Kamaruddin
Selangor, Malaysia — Nur Izzah Nabilah Haris
Selangor, Malaysia — Paridah Md Tahir
November 2024

Contents

Wood and Non-wood Engineered Products: An Overview 1
Dishari Chatterjee, Souvik Ray, and Sanmitra Dan

Lignin-Based Wood Adhesives 29
Ega Cyntia Watumlawar and Byung-Dae Park

Binderless Lignocellulosic Composites for Green Material 67
Juliana Abdul Halip, Aisyah Humaira Alias, and Balkis Fatomer A. Bakar

Lignin-Based Adhesive Bonded Veneer Composites: A Sustainable Approach 89
Shibu Comath, Arya Chirakkal, and Ashline Shaji

Particleboard and Fiberboard Bonded with Green Binder 111
Nur Izzah Nabilah Haris, Syeed SaifulAzry Osman Al Edrus, Nur Haninah Harun, Arif Azizi Che Harun, Nurain Johar, and Mohd Hafizz Wondi

Performance of Wood Polymer Composite Using Eco-Friendly Binders 125
Siti Norasmah Surip, Falah Abu, Mohd. Nazarudin Zakaria, Siti Rafedah Abdul Karim, and Azrena Abdul Karim

Potential of Green Binders as Matrices in Composite Materials 141
Fatin Nur Azmina Mohd Fauzi, Azra Umairah Anuar, Noor Najmi Bonnia, Nur Izzah Atirah Mat Hussin, and Aein Afina Redzuan

Development of Recycled Cotton with Corn Starch as Binder for Sound and Thermal Insulation Panel 165
Tasnim Zakepeli, Kamarul Aini Mohd Sari, and Fatimah Mohamed Yusop

Polyurethane from Liquefied Bamboo as Green Wood Coating Materials 175
Redzuan Mohammad Suffian James, Paridah Md Tahir, and Norwahyuni Mohd Yusof

Environmental and Economic Impacts of Green Panels 191
Nur Izzah Nabilah Haris, Syeed SaifulAzry Osman Al Edrus, Norfaryanti Kamaruddin, Nurliyana Abdul Raof, A. W. Noraida, and Mohd Hafizz Wondi

Wood and Non-wood Engineered Products: An Overview

Dishari Chatterjee, Souvik Ray, and Sanmitra Dan

Abstract Wood is one of the most abundantly found sustainable materials in the world that serves a wide range of purposes from daily household items to structural purposes. The other lignocellulosic materials of lower qualities and strength can be modified into boards using a variety of processes and glues. The strength of these boards largely depends on the adhesive used, and the process used to manufacture them. In the past two decades, there has been a surge in the engineered wood sector, and it is mainly due to the increasing concern about carbon sequestration, sustainability, and renewability. This chapter focuses on the different types of formaldehyde-based and biobased glues used in the production of lignocellulosic composites. The review also focuses on the various types of engineered wood products available in the market for structural and nonstructural applications. The chapter thus provides key aspects of sustainable development by focusing on the new biobased technologies and non-wood composites to develop a greener future.

Keywords Engineered wood · Lignocellulosic composites · Biobased · Adhesives · Sustainable

1 Introduction

A growing environmental consciousness has made the production, usage, and replacement of conventional building materials (polymers derived from petroleum, glass fibers, and other products) a priority. The United Nations Conference on Trade and Development (UNCTAD), March 2023 report showed that in the second half of 2022 commerce in environmentally friendly products increased by roughly 4%. The combined worth increased by more than \$100 billion from 2021 to a record \$1.9

D. Chatterjee · S. Ray (✉) · S. Dan
Wood Properties and Processing Division, Institute of Wood Science and Technology, Bangalore 560003, India
e-mail: souvik.ray@lnu.se

S. Ray
Department of Forestry and Wood Technology, Linnaeus University, Växjö 35195, Sweden

S. S. Osman Al Edrus et al. (eds.), *Green Lignocellulosic-Based Panels*, Engineering Materials, https://doi.org/10.1007/978-981-96-3404-0_1

trillion in 2022. To offer housing that is ecologically friendly while also providing for the necessary occupant environment and operating efficiency, designers and builders are being challenged by today's housing needs to use more imaginative materials.

There is a need to examine the state of knowledge and projected advancements in the field of bio-based composite technologies to address this challenge [99]. Among several categories identified for bio-based descriptions and of importance to the housing industry, we will be discussing here the wood and non-wood-based engineered products. Talking about wood-based engineered products, wood is the main constituent which accounts for 94% or more of the mass [14]. Wood-based panels are one of the widely used categories of wood-based engineered products. Veneers, wood strands, wood particles, and wood fibers are typical components of traditional wood-based panel goods. In contrast, non-wood-based engineered products are mostly made of non-wood lignocellulosic substances and leftovers from agriculture. Composite non-wood panels are created by carefully blending items made from recycled paper or agricultural residue with bio-based adhesives and coloring agents [21].

Panel products made of wood have a variety of structural and nonstructural applications. Product categories include furniture parts, indoor and outdoor panels, and building support structures. Wood-based composites are made from a wide range of materials, such as timber and fibers from underutilized or small-diameter plantation trees. Wood-based panel items provide consistent and predictable in-service outcomes regardless of the materials that are used in their production, primarily because of the guidelines that are used to track and govern their production. The mechanical properties of wood-based panel products are determined by the density of the finished product, the type of adhesive used to bind the wood elements together, the geometry of the wood elements (fibers, flakes, strands, particles, veneer, lumber), the category of forest management Systems (e.g., naturally regenerated, intensively managed), and the type of wood elements [13, 14].

However, contemporary international and regional initiatives, such as the Sustainable Development Goals (SDGs) of the United Nations are given top priority for the upholding of the conservation of biodiversity, and forests both in terms of quantity and quality [72]. The wood-based engineered panel sector is compelled to explore other raw material sources due to this, added to rising timber prices, and transportation issues brought on by hampered logistics. The Sustainable Development Goals (SDGs) have significant impacts on the wood composite industries. The industry is increasingly using composites from renewable and sustainable resources. Governments, regulatory organizations, and consumers are demanding materials and process solutions that favor decarbonization [33]. Voluntary sustainability standards can offer opportunities for developing and developed countries alike to access new markets [111]. However, studies on the compatibility of recycled and waste materials with other components in composite structures for improved interface and better mechanical performance pose major scientific challenges [71].

The SDGs are influencing the wood composite industries to innovate and adopt more sustainable practices, which in turn can lead to new market opportunities and regulatory benefits. The industry for Wood and Non-Wood Engineered Products is going to be directly shaped by the following 5 SDGs, among the 17 SDGs, which were established by the UN in 2015: SDG 9 (Industrialization, Infrastructure and Innovation), SDG 12 (Responsible Consumption and Production), SDG 13 (Climate Action), SDG 15 (Life on Land), and SDG 17 (Partnership for the Goals) [119].

Non-wood lignocellulosic material represents a feasible alternative supply of essential elements for the woodworking industry [92]. Since they come from renewable resources and are readily accessible, lignocellulosic materials other than wood can be potential alternatives for raw materials for the wood sector. While their density is often lower, non-wood lignocellulosic materials have several traits in common with wood, such as anisotropy and hygroscopicity, for making low-density composites [73]. Because of their low mechanical properties and high concentration of fines, wooden fragments or fibers cannot be totally replaced with substitute materials [125].

As a result, using substitute materials in manufacturing engineered panel products is limited. If not, more glue would be needed to provide the panels with the necessary level of mechanical strength [81]. The fibers that serve as reinforcement for both wood and non-wood panel products are bound together by a polymer matrix to create a material with the requisite rigidity and strength for structural use [21]. Figure 1 represents the classification of wood and non-wood biofibers.

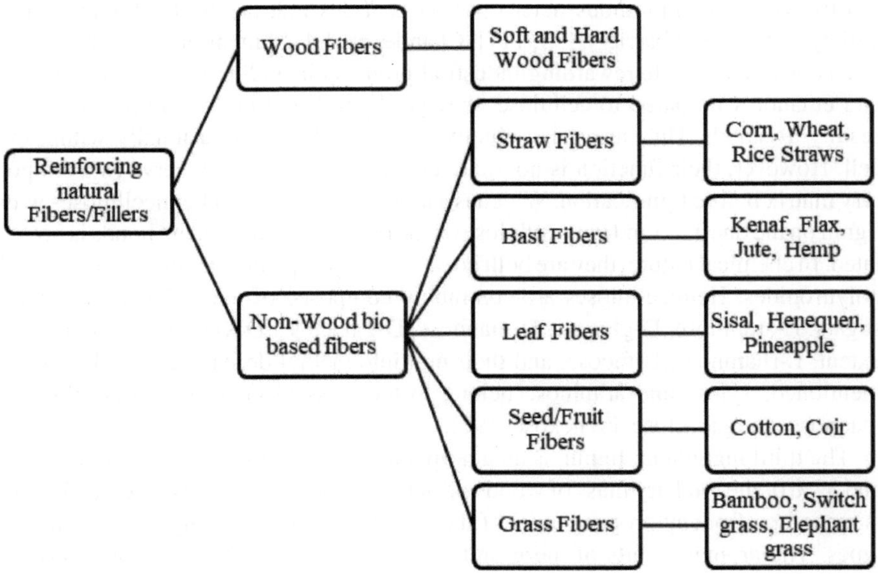

Fig. 1 Classification of different kinds of natural reinforcing fibers adopted from Drzal et al. [21], open access

2 Chemical Properties of Lignocellulosic Materials

The primary components of wood are cellulose, hemicellulose, and lignin. A framework made of aromatic lignin encases the cellulose and hemicellulose that constitute wood [78]. Composed of glucose subunits and connected by β (1 → 4)-glycosidic bonds, cellulose is a linear polysaccharide with a high degree of polymerization. The straight and stiff cellulose chain is a result of the glycosidic bond's β-configuration. While lignin serves as a binding agent between the fibers, the cellulose and hemicelluloses dictate the mechanical properties of wood fibers.

Cellulose chains are packed together into bundles called microfibrils. Microfibrils are further bundled together to form cellulose fibers. The fibers are embedded in a matrix of other cell wall components, such as hemicellulose and lignin. This composite structure gives plant cell walls their strength and rigidity. Because of the staggered configuration of the glucose units in cellulose, hydrogen bonds can form between adjacent chains. The hydrogen bonds form a strong network that holds the chains together and prevents them from dissolving in water. In addition to hydrogen bonding, there are also van der Waals forces between the cellulose chains. Van der Waals forces also contribute to the strength and rigidity of cellulose.

The second most prevalent polysaccharide in nature, hemicelluloses [109] are mostly found as amorphous hetero polysaccharides that are useful for making a variety of films and barriers. [35, 117]. Chandel et al. [15] mentioned it can be used as a substrate to create rewarding industrial products including butane-diol, xylitol, and ethanol. Compared to cellulose, they get hydrolyzed more readily, but not as readily as starch. They are found chiefly in cell walls and occasionally within the cell. However, their function is not quite clear, presumably, they serve as a temporary matrix before lignification. Which is a loose combination of hemicelluloses with lignin on the one hand and with cellulose on the other through covalent bonds is postulated. In chemical nature, they are both polysaccharides (pentosans and hexosans) and polyuronides. Hemicelluloses are commonly composed of the following common sugars: L-arabinose, D-glucose, D-mannose, D-galactose, D-xylose, and, to a lesser extent, L-rhamnose, L-fucose, and their multiple methyl derivatives. Of the sugars mentioned, xylose and arabinose belong to the class of pentoses, while glucose, mannose, and galactose are hexoses [96].

The third ingredient, lignin, is an amorphous polymer that makes up between 20 and 32% of the total dry mass of wood and acts to cement wood cells together. Lignin is made up of phenylpropane units of the p-hydroxyphenyl, syringyl, and guaiacyl types. The aromatic units of these units have one, two, or three free or etherified hydroxyl groups. Monolignols in the plant cell wall matrix undergo oxidative polymerization to create lignin [68]. Alkyl aryl ether (α-O-4, β-O-4) and aromatic ether (4-O-5') links, as well as carbon–carbon bonds (5–5' or β–5), bind the phenylpropane units together in condensed states. Alkyl aryl ethers are classes of organic compounds that contain an alkyl (radical alkane) group bonded to an aryl (or aromatic) group through an oxygen atom. Because of its phenolic resemblance, lignin has been proposed as a formaldehyde-based resin substitute [97].

Lignin is positioned in such a way that it forms an encapsulating material in and around the carbohydrate fraction which makes it easily accessible for modification. Structurally, lignin is a three dimensional optically inactive polymer formed by the dehydrogenative polymerization of three p-hydroxycinnamyl alcohol precursor molecules coumaryl alcohol, coniferyl alcohol, and sinapyl alcohol linked together in an irregular manner. Dehydrogenation refers to the removal of hydrogen from an organic entity such that radical formation takes place. Polymerization via these hydrogen-lacking radicals refers to dehydrogenative polymerization. Lignin is insoluble and hygroscopic in its natural state. However, when lignin undergoes hydrolysis, it breaks up into smaller structural units and becomes soluble in water.

3 Adhesives

The adhesives used in wood are polymeric substances that can either react physically or chemically or in both ways, with the surface of the wood to transfer stresses between bound members in a safer manner. Contemporarily, the adhesives and the physicochemical processes of adhesion are involved in more than 70% of all wood-based materials. In addition to meeting the requirements of the bonded assembly for its intended application, an adhesive's properties must also be compatible with the substrate and bonding conditions. Either synthetic or bio-based adhesives are typically utilized in the industry for wood and panel adhesion.

Synthetic adhesives like melamine formaldehyde (MF), phenol formaldehyde (PF), and urea formaldehyde (UF) have formaldehyde as one of the constituents and are popular in the market due to their availability and affordability [20]. PF is the most popular amongst all due to its superior water resistance and bonding ability. Recent research has now been focussed on polyurethane, isocyanate-based, polyvinyl acetate, and epoxy-based adhesives, because of strict health and environmental regulations [121]. A few investigations on bio-based adhesives, such as protein isolate-modified, soy-based, and methylated black liquor adhesives, have recently been conducted [104]. However, compared to conventional adhesives, their bonding strength is weaker [84].

3.1 Synthetic Adhesives

Polymers made from petrochemically generated elements (organic compounds) generally make up synthetic adhesives. Spreading and moistening the wood's surface requires the application of adhesives in the liquid state. Synthetic adhesives can further be classified based on their nature of curing as thermoplastic or thermosetting adhesives. Thermoplastic adhesives are those adhesives which melt on being exposed to heat and re-solidifies on cooling. A major example of this category is the PVAc adhesives. Thermosetting adhesives on the other hand are insoluble and are irreversibly cured or solidified via cross-linking of the polymers. They do not soften on further heating. A major example of this kind are the phenolic resins.

Formaldehyde-based adhesives

The most widely used category of wood adhesives is derived from formaldehyde reacting with phenol, urea, resorcinol, melamine, or a combination of any of the two compounds. Typically, formaldehyde-based resins used are water-borne in nature, which means the curing process undergoes a polymerization reaction, accompanied by the loss of the solvent i.e. water [27]. The majority of laminates, finger joints, and composite goods are using formaldehyde copolymer adhesives, while isocyanates are gaining market share due to their decreased sensitivity to wood moisture and processing temperatures. Because formaldehyde forms crosslinking groups in addition to polymeric chains, these adhesives provide rigid, non-creeping bonds and robust wood adherence. But the characteristics differ according to the co-actant utilized with the formaldehyde.

Phenol formaldehyde adhesives

The oldest class of synthetic polymers, phenol formaldehyde (PF) polymers, were prepared at the commencement of the 20th century. Because of their outstanding durability, these resins are widely used in laminations and composites, which is attributed to their strong adherence to wood, high polymer strength, and great adhesive stability. In most of the durability tests the occurrence of wood failure is much higher than delamination for PF adhesives. There is a plethora of formulations that might be used, and choosing the incorrect one could result in weak bond strength. PF adhesives are manufactured by reacting phenol with formaldehyde such that can go through further polymerization during the setting phase. Pre-polymers can be divided into two categories: novolaks, which are typically produced under acidic circumstances (pH range of 4–7) and have formaldehyde/phenol (F/P) ratios less than 1; while resole resins, which are often produced under basic conditions (pH range of 7–13) and have F/P ratios higher than 1.

Resorcinol and phenol-resorcinol formaldehyde adhesives

Resorcinol-formaldehyde (RF) is a thermosetting class of adhesive, RF has two -OH groups at -1 and -3 of the aromatic rings making it highly electrophilic in nature and it can be cured at room temperature also. Due to the existence of three reactive sites, phenol and resorcinol can readily demonstrate cross-linking to produce a thermosetting polymer. RF resin is used to produce a solid primer of low molecular weight, known as hydroxy-methylated resorcinol (HMR). RF resins form very strong and durable bonds like PF; however, the higher cost of resorcinol is the main drawback of this resin. Thus, to regulate the cost and to maintain the room temperature curing of the adhesive, the mixture of phenol, resorcinol, and formaldehyde (PRF) was introduced. PRF adhesives are extensively used in finger joints and laminations [27].

Urea formaldehyde adhesives

Urea-formaldehyde (UF) is a non-transparent, white-coloured thermosetting adhesive produced from the mixture of urea and formaldehyde. Plywood, particle board, medium-density fiberboard (MDF) and moulded objects are manufactured using UF. These resins exhibit high flexural modulus and tensile strength and resistance against high heat-distortion temperature, thus making them very popular as a wood glue as well [27]. Generally, hardeners like NH_4Cl are used for the fast curing of UF adhesives. UF resins are generally cured at 110 °C. The UF resin is popularly used mainly due to its cheaper cost as compared to phenol and resorcinol adhesives, however, it suffers from various drawbacks mainly, a higher rate of formaldehyde emissions and poor water resistance.

Melamine formaldehyde adhesives

Like UF, Melamine-formaldehyde condensation produces Melamine-formaldehyde (MF) adhesive but compared to UF, MF adhesives have better water resistance. The most popular applications for MF resins are finger joints, exterior and semi-exterior plywood, and particleboard. The major drawback of MF adhesives is their exorbitant price. The adoption of melamine-urea-formaldehyde (MUF) resins, which have higher water resistance than MF resins but are significantly less expensive, has resulted from this. The formaldehyde-to-melamine ratio in typical applications ranges from 2 to 3. A typical MF resin for wood bonding has between 53 and 55% solids, and its pH ranges from 9.9 to 10.3. It is common practise to add a hardener, such as ammonium chloride or sulphate, which will produce ammonia migration from the adhesive as well as hydrogen chloride or hydrogen sulphate [27].

Isocyanates-based adhesives

Isocyanates can self-polymerize or interact with numerous different monomers, thereby giving producers a tremendous deal of versatility in the goods they may create. Typically, polyurethanes are made from isocyanates by reacting with liquid diols. Isocyanates tend to react readily, there are advantages and disadvantages to isocyanate's strong reactivity. Isocyanate which can cure by itself, is the most popular type of wood glue that reacts with the water inside the wood and gets cured. The polymeric diphenylmethane diisocyanate (pMDI) is the only known effective adhesive for strawboard binding and is mostly utilized for the manufacturing of oriented strandboard cores [27]. Due to the quick cure ability of pMDI, the market trend is shifting from phenol-formaldehyde adhesives. Isocyanates have been widely employed in coating applications; however, non-waterborne two-component isocyanates are rarely used as wood adhesives due to their quick response times.

The curing process is initiated by mixing the diisocyanate with a diol, resulting in a linear polymer, typically, with a modest degree of crosslinking to produce more flexible goods. The polymer backbone of hot melt adhesives is generally attached to isocyanate groups. As the liquid polymer cools and changes from a melt to a solid, hot-melt adhesives gain binding strength, which makes them particularly attractive for product assembly. Sadly, the bond typically has weak resilience to heat, long-term stressors, and occasionally wetness. Polyurethanes, the other family of isocyanate adhesives, are utilized in more specialized wood-bonding applications. The potential for polyurethanes to bond green wood has been quite promising.

Epoxy adhesives

Most wood adhesives need heat cure, and adherence to a wide range of substrates is weak. Epoxy adhesives address these issues since they cure at room temperature, fill in gaps, and adhere to other surfaces. While some epoxies can self-polymerize in the presence of acid or tertiary amine catalysts, most epoxies have a strongly crosslinked alternating ABABAB backbone that often takes advantage of the hardener's (popularly amines, but also thiols, hydroxides, and acid groups) multiple functions. Epoxy resin consists of mainly two components: bisphenol A and epichlorohydrin. Epoxy adhesives have extensive utilization due to their excellent environmental resistance and capacity to adhere to a wide range of surfaces [27]. However, due to high cost, and in some cases lower durability they are less frequently employed in wood bonding. However, there is some dispute regarding the resilience of epoxy bonding in moist environments.

Polyvinyl and ethylene-vinyl acetate dispersion adhesives

Polyethylene-vinyl acetate and polyvinyl acetate are water-borne adhesives, popularly used in wood assembly and in paper parts of finished goods. For wood bonding, common white glue (PVA) dispersions are primarily utilized. The pre-formed polymers do not need heat for curing, and are cost-effective and simple to use. In comparison to the hard formaldehyde copolymers typically employed as wood adhesives, PVAs are aliphatic-backed polymers with linear structures, making them far more flexible adhesives. Usually, these PVA adhesives flow freely into the exposed cell lumens but are unlikely to penetrate cell walls due to their large molecular weight. PVAs fail to perform well because of their poor creep resistance which leads to strength loss under high levels of continual stress under high moisture levels [27]. Cross-linked PVAs are more resilient to conditions with elevated moisture content temperatures, and, stresses but they are less useful because the cross-linker needs to be applied immediately before application. They can be utilized in other situations, such as the building of windows and doors, when ordinary PVA is ineffective.

3.2 Biobased Adhesives

The early 1900s saw widespread use of bio-based adhesives in the bonded wood business, which grew significantly as more efficient adhesives were produced. Any kind of adhesives of natural origin such as albumin, glutaraldehyde, chitosan, cyanoacrylate, fibrin and thrombin, gelatine, polyethylene glycol, and urethanes are referred to as bio-based adhesives, protein-based adhesives are being most popular among them. The biggest development was the creation of adhesives made from soybean flour, which made interior plywood a more affordable alternative to solid wood.

Many adhesive compounds are naturally found in trees and plants, some of which have been applied to the joining of wood. The protein-based glues, tannin adhesives, and lignin adhesives are some of the significant bio-based adhesives. Tannin and lignin have been intensively researched as phenol replacements in phenol formaldehyde resins.

Protein-based adhesives

Protein-based adhesives have been derived from numerous sources, including fish skins, soybeans, milk (casein), bones, and hides from animals. The once-prevalent protein glues used for wood bonding, are currently employed in a specialized field where they offer some unique features and are still being researched for usage in limited durability wood bonding. Proteins, like most biomass materials, don't have a constant composition, so the methods for using them and the characteristics of the adhesives, change as well. However, the polyamide backbone forms the fundamental structure because of the amino acid's condensation. As hydrogen bonds are broken during denaturization, various secondary and tertiary bonds may also be broken, depending on the circumstances [27]. The denaturation of the protein causes it to flow into the wood and form hydrogen bonds with the wood components. Soybean is utilized in the greatest amount of protein-based adhesives. Other adhesives used are blood protein extracted from beef and hogs which offers the best performance in terms of water resistance. Fine furniture requires flexible as well as durable bonding in variable moisture conditions of indoor environment that is why animal bone and hide glues are used for this purpose. Casein is utilized in fire doors because, like many protein adhesives, it offers good fire resistance. Each of these adhesives has its own method for application and denaturization [27].

Tannin-based adhesives

Tannins are polyhydroxy polyphenolics that are present in a wide variety of plant species. Tannins are employed despite being costlier than phenol because they are more reactive than phenol. Formaldehyde mostly added as single additions, but it can also occasionally add two times to the linked rings of the tannin structures of pyrogallol, phloroglucinol, catechol, and resorcinol. The final polymer structure is significantly different from the resorcinol product because of its lower crosslink density [27]. When compared to synthetic adhesives, tannins have three drawbacks: high viscosity, restricted access, and unpredictable source, which affect reactivity. The polycyclic structure of the tannins contributes to its high viscosity. Tannins are present in a few species at concentrations that make them commercially feasible. Because tannins' composition fluctuates depending on the growing environment, it is challenging to produce adhesives that work consistently [27].

Lignin-based adhesives

Despite being phenolic derivatives, lignins are very different from tannins, as they can be easily procured and less expensive, they also exhibit slow reactivity with formaldehyde. Lignin is abundant because it is a by-product of the pulping operations used to make paper making up to 24–33% in hardwoods and 16–24% in softwoods.

The crosslinked polymers in native lignin are first partially degraded to isolate from the complex matrix. This degraded lignin is then polymerized to acquire beneficial sticky qualities. Lignin has limited formaldehyde reactivity because of its lesser number of phenolic rings and lack of polyhydroxy phenyl rings despite being virtually entirely aromatic [27]. However, the lignosulfonates in the wasted sulphate liquor (SSL) obtained from the sulphite pulping process of wood are a more useful feed for the generation of reactive lignin [27].

Curing techniques, such as acid curing at elevated temperatures and coupling in the oxidative environment using hydrogen peroxide and catalysts, increase reactivity with formaldehyde. There are three ways to use SSL as the primary adhesive: extended press periods with a post-heating phase; sulphuric acid assisted bonding heating; and heating using hydrogen peroxide. Furthermore, SSL has been used as extenders for urea and phenol formaldehyde resin. Pre-methylation with formaldehyde changes the low reactivity of lignin, and pre-methylated lignin can be used in plywood bonding when combined with PF resins [27].

4 Engineered Wood Products

Engineered wood products (EWPs) are created from hardwoods and softwoods, singularly, or in combination of both, and joined together with adhesives of suitable choice. These items are treated on a regular basis to increase their quality and capacity. EWPs are a diverse group of product types with a wide range of production methods and uses. Engineered wood products include particleboard, plywood, fiberboard, and mass timber constructions (MTCs) like oriented strand board (OSB), laminated veneer lumber (LVL), glue laminated timber (GLT), and cross-laminated timber (CLT). Plywood, particleboard, and fiberboard manufacturing rise yearly, with a reflected volume of almost 380 million m^3 in 2021. [45]. With a production of around 167 million m^3, plywood is the largest amount produced, followed by fiberboard (about 110 million m^3) and particleboard (about 100 million m^3).

Plywood

One of the most well-known multipurpose engineered wood-based panel products is plywood. It is basically a composite material made from layers of thin wood veneers glued together by a suitable adhesive with the alignment of grain direction of two adjacent veneers perpendicular (90°) to each other. The top face of plywood is made up of face veneers, the bottom face is made up of veneers known as back veneers and the central layers of veneers are known as cross back. Generally, both hardwoods and softwoods are used for the manufacture of plywood and the face veneer is always of superior quality than the cross backs. Veneer is the chief ingredient for the manufacture of plywood. The term veneer describes very thin wood slices, usually not more than 3 mm (1/8 in.) thick, that are attached to core panels. The core panels are usually made of wood, such as particleboard, or medium density fiberboard. Veneers are also used to create flat panels for furniture pieces, doors, tops, and cabinet panels. Veneer is extracted by slicing big, rectangular blocks of wood called flitches, or by peeling the trunk of a tree. Cutting by slicing through a tree's growth rings gives wood its grain and figure, which vary depending on the angle at which the wood is cut. Using almost the entire log, veneering is a very common method of highlighting the natural characteristics of wood.

Plywood has gained much importance and popularity mainly because of its advanced properties and the wide range of advantages that it provides over solid wood. Plywood is made up of veneers, which makes it lighter than solid wood products, thus providing ease of operation. The direction of fibres of the adjacent veneers in plywood is placed orthogonal to each other. These orthogonally placed fibers provide greater strength to the plywood and make it more resistant to shrinkage and swelling on exposure to moisture and humidity. In terms of the four fundamental engineering strength attributes of bending, tension, compression, and planar shear, plywood is more robust than alternative wood-based panels. Another major benefit of plywood is that plywood can be manufactured from fast growing plantation timbers which still gives good strength properties as compared to those of solid wood furniture. Also, since the individual veneers are sliced very thin, there is almost zero wastage as the defects, if present, are equally distributed throughout and not in a single place. Thus, the loss in strength properties is also minimal.

Particle boards and fiber boards

An engineered wood product comprised of wood chips is called particleboard. Particle boards are thermally pressed after mixing with synthetic resin or another appropriate binder. Particleboard is one of the widely spread markets in the wood-based goods traded internationally. Both worldwide production and demand for it have increased recently. In 2020, the worldwide particleboard market was valued at 21 billion US dollars, according to IMARC Group [63]. The group projects a 4.4% compound annual growth rate (CAGR) for the following six years. Europe, the Americas, Africa, and Oceania are the regions that create the most particleboard, with Germany, Poland, Italy, Austria, and France being among the top producers. Asia lags behind these regions in terms of particleboard production [63].

Wood and non-wood biomass are two categories of agricultural biomass. Wood biomass consists of both hardwood and softwood species. According to [6], leaf, grass, seed, fruit, stem, straw, and stalk segments can be used to categorize non-wood biomass. The wood species commonly used for particleboard productions are lesser known, underutilized timbers which are generally light coloured woods [40, 44, 62]. The rubber wood (*Hevea brasiliensis*) and sengon wood (*Paraserianthes falcataria*) in Southeast Asia, are the most used timber species for particleboard manufacturing.

The raw material used for non-wood particleboard preparation varies widely ranging from tobacco [2], cotton [80], sunflower [30], sorghum [57], to chili pepper [87], rapeseed [86], tomato [113], kenaf [122], and cup-plant [59] plant stalks. However, commonly used non-wood biomass are wheat straw [32], rice straw [64] and sugarcane bagasse [112]. The studies showed that the wheat straw particles reinforced with UF board properties were inferior to those commercially available ones, but the rice straw boards and sugarcane bagasse boards performed satisfactorily and even better in the case of eucalyptus and pine counterparts [69].

Fiber is a polymer that is procured through agricultural natural resources and is used to manufacture bio-composite boards. Clarifying more, [4] mentioned any agricultural wastes like wheat, rice and corn all fall under the category of fibers. Straw fiber bio-composite boards are developed by combining two or more fibrous materials to create a new product that has specific, unique characteristics and is typically more useful and functional. Due to their negligible effect on the environment, using materials of natural origin has been continuously observed in recent times. The manufacturing of composites from renewable resources is expected to rise from 12% in 2010 to 25% in 2030. Biocomposites reinforced with natural fibers have found widespread application in the packaging, construction, aerospace, and automotive industries, where high load bearing capacities are not necessary [128]. Natural fibers provide the advantages of being lightweight, having a high specific strength, being easily formed, and having a significant resistance to corrosion and fatigue [95] but the drawbacks include more moisture absorption and a high degree of anisotropy [50].

Fibers that can be used commonly are wheat straw fibers, rice fibers and corn corb fibers along with other wood fibers. Fibers from wheat straws is one of the most readily used raw materials in fiberboard preparation. The mechanical properties of the conventional textile bast fibers are superior to the wheat fiber bundles. Wheat fiber has a low bending endurance, which can be overcome by reconstituting the fibers which might enhance the properties of its product, but it is only practical to use it in whole stem form [107]. A comparative study of straw fibers revealed that the mechanical characteristics of composites made of wheat straw fiber outperformed composites made of corncob and cornstalk fibers [90]. Research of [3, 9] have revealed the potential use of wheat fibers as filler in a poly-3-hydroxybutyrate-co-valerate (PHBV) biopolymer and polycaprolactone and wheat fibers [36]. By employing a torrefaction process, wheat straw fibers' hydrophobicity can be enhanced, and melt-extrusion can be used to create bio-composite boards. In contrast to untreated fibers, matrix polymers' molecular weight and crystallinity are retained intact [105]. Due to the glassy gluten matrix's qualities being like those of an epoxy resin, the utilization of wheat fiber gluten has received increased interest recently [41]. The addition of wood powder to non-wood boards is found to improve the properties substantially. 60% pine particles were added to sugarcane bagasse board to achieve the optimum mechanical and physical qualities when the board was bonded with polyurethane (PU) resin based on castor oil [12]. Similar observations were also observed when researchers incorporated 50% industrial timber residue into sugarcane bagasse particleboard [124] and perennial ryegrass with *Eucalyptus camaldulensis Dehn.* wood particles [82]. Particleboard preparation has also been attempted with sunflower hulls [29], walnut (*Juglans regia* L.) shell and almond (*Prunus amygdalus* L.) shell bonded with UF resin [94], hornbeam (*Carpinus betulus* L.) and beech (*Fagus orientalis* L.) wood, waste tea leaves (*Camelia sinensis* L.) bonded with UF resin [123], perennial ryegrass (*Lolium perenne* L.), reed canary grass (*Phalaris arundinacea*), jose tall wheatgrass (*Agropyron elongatum*), vetiver grass (*Chrysopogon zizanioides*) and bamboo [63]. The boards prepared from waste tea leaves are found to be exceptionally resistant to brown-rot fungus (*Tyromyces palustris* (Berk. et Curt) Murr.), white-rot fungus (*Coriolus versicolor* (L. ex Fr.) Quel.) and subterranean termites (*Coptotermes formosanus* Shiraki) [123].

Pine needles have always been a hot topic when it comes to the preparation of particle boards. When different combinations were experimented, the mechanical and physical characteristics of the thermally treated recycled pine particles exhibited small differences [7]. In comparison to other samples, composites made from pine particles treated at 220 °C exhibited a notable decrease in hygroscopicity and this was in accordance with observations of [91]. Further investigations were conducted with several kinds of liquefied wood waste binders and recycled wood particleboard [49].

Enzyme-based composites

Laccase EC 1.10.3.2, p-diphenol dioxygen oxidoreductase, is part of the multicopper enzymatic group that catalyzes the oxidation of phenolic compounds, reducing oxygen into water. The oxidized reactive aromatic molecule enhances the activity of the lignin part and provides ideal surface for grafting through covalent bonds [85, 89]. Laccase acts as an oxido-reductase agent; on one side it removes the lignin molecule from cellulosic material and on the other side it helps in vito polymerization of lignin through free radical reactions [127]. It can degrade lignin without lignin peroxidase and manganese peroxidase. Widsten and Kandelbauer [118] summarized the adhesion improvement of lignocellulosic products, such as medium-density fiberboard and particleboard, by enzymatic bonding methods. This may involve two approaches, one is to improve the self-bonding properties of the particles by oxidizing their surface lignin before being constructed into boards. Another technique involves utilizing enzymatically pre-treated lignin as adhesives for boards and laminates.

The first pilot-scale production of laccase-assisted fiberboards was made from fibers of beech (*Fagus sylvatica*) and was reported by [25]. Although the strength properties of the enzyme-bonded boards were analogous to those bonded by UF-resin and wax, however, the dimensional stability of the enzyme-bonded boards was not sufficient. In enzyme-bonded boards, a cross-linking of the lignin was monitored but no conclusive result for whether the cross-linking formed covalent bonds between fibers [26] or between loosely associated lignin worked as an adhesive. However, the nature of laccase-catalyzed bonding requires more pressing temperatures and times, and thus concerns economic feasibility and sustainability. Thus, the biggest problem in promoting ecofriendly bio-based panels and fiberboard is the low rate of industrial acceptance.

However, tremendous improvement and standardization have increased laccase's utility as an eco-agent for both polymerization and grafting [97]. Organo-solvents in polymerized state can be used as the bonding agent, and the use of copolymers could enhance their connections even further. Finally, laccase-mediated polymerized lignosulfonate may be useful as a water-based primer for wood surfaces. In nature, laccase reacts with the lignin part of the cell component to generate free radicals of lignin, and distinct lipophilic extractives [100]. Medium Density Fiber (MDF) boards prepared from laccase-treated rubber wood (*Hevea brasiliensis*) and organosolv lignin exhibited higher physical, mechanical, and chemical properties. Organosolv and con-organosolv were used to manufacture the boards, with a target density of 810 kg/m^3 [75]. The physiochemical characteristics of enzymatically treated rubber wood medium-density fiberboard (MDF) were found to be superior in terms of fiber surface structure and thermal deterioration when compared to untreated fiber boards [79]. The removal of extracellular lignin from the fiber surface resulted in a higher crystallinity index. The thermo-gravimetric analysis demonstrated that treated fibers began deteriorating at higher temperatures than untreated samples. Furthermore, the treated fiberboards demonstrated improved dimensional stability [77]. The boards that had been treated with enzyme for a longer period had the slightest swelling. This is due to the limited access to fiber due to the absence of any free hydroxyl group.

The fiberboard with a density of 750 $\pm$ 10 MPa was made from fiber that was treated at 9 U/g for 60 min and pressed at 200 °C for 6 min. The binder less fiberboards from laccase-treated rubber wood fiber and demonstrated enhanced performance when evaluated according to ASTM D1037 [76]. Laccase enzyme oxidized lignin to form a stable phenoxy radical, enabling another component to adhere to it; thus, acting as a potential cross-linking agent. Laccase-treated rubber wood fibers were employed to manufacture MDF with Enzyme Hydrolysis Lignin (EHL) and Concentrated EHL. Con- EHL had a lower stretching energy in the benzene region than EHL and a curing behavior comparable to urea formaldehyde [74]. The prepared MDF boards had improved mechanical strength and passed the ASTM D1037 internal bonding and modulus of rupture standards. The enhanced properties of laccase-treated fiber and Con-EHL demonstrated a viable option for preparing a new, eco-friendly MDF board.

Preparation of lignosulfonate-based adhesive via polymerization using laccase enzyme was also achieved [37]. The polymerized lignosulfonate adhesive (LSp) had higher viscosity when polymerized for a prolonged period. The mechanical test results of samples treated with LSp were observed to be comparable to samples treated with PVAc glue. While the samples with PVAc were found to have wood failure, the samples with PVAc were mainly glued-line failure.

Fungal-based composites

The concept of composites fused by fungal mycelium is relatively novel but has received significant consideration from both academia and industries. The most considered and marketed mycelium-based products are in situ-mycelium-based composites [23, 52]. Typically, the production of fungal boards starts with inexpensive lignocellulosic biomass substrate onto which mycelia of filamentous fungi are added, such that the mycelium partially processes and sticks to the surface of the substrate materials. As the fungi grow, the mycelium forms entwined networks and bind the substrate into a cluster. Oven drying is a common method used in both industry and academic labs to destroy fungal growth and thereby generate a finished product [23].

Mycelium-based bio-composites, or lignocellulose-based boards bonded by fungus, are a relatively new idea but have garnered substantial academic and commercial interest. Usually, a low-cost lignocellulosic biomass serves as the foundation for their fabrication on to which filamentous fungi's mycelia are put in with nutrients and substrate with a specified form, where the substrate particles are partially digested and get attached by the mycelium. The binding of substrate particles into a single aggregation is done by the entangled networks of the mycelia. The final product is then prepared by inactivating the fungi under an autoclave [23] and then pressing the boards as per requirements [52].

In biomass substrate, the feeding behaviour of the white-rot fungus results in reductions in mass, mechanical properties, and density of the substrate along with an increase in porosity and water absorption [10, 126]. The mycelium matrix governed primary compressive behaviour at minor strains (<20%), whereas the lignocellulosic substrate oversaw stiffening at greater stresses, [47, 48]. Bruscato et al. [11] found that the fungal mycelia's properties influence the compressive strength, high compressive strength in the final composite is associated with leathery mycelia, as opposed to brittle mycelia. In general, the mycelium's bottom surface forms a stronger attachment than its top surface because it is denser, flatter, and more hydrophilic in nature. Thus, the surface mycelium layer might be used as an independent adhesive for binding untreated wood. [110].

Mass timbers constructions

Wood is the most often utilized renewable resource and one of the earliest building materials used by humans. However, after industrialization, the emphasis moved to cement and concrete as building materials. Nowadays, green energy and sustainability are the primary goals to save our planet from environmental damage, as large amounts of energy are necessary to create cement and concrete. Furthermore, their manufacture causes significant environmental damage, elevating the concentration of carbon in the earth's atmosphere, which is detrimental to humans. These construction materials are very energy and carbon-intensive thereby causing environmental pollution. The demand of wood for residential and non-residential construction has increased because of its renewability and less energy intensiveness.

Mass Timber Construction (MTC) is a term used to describe a family of major engineered wood products that have structural applications in construction and have been marketed as building industry alternatives to concrete and steel. Wood-based materials have achieved significant developments in exterior non-structural or semi-structural construction materials such as the frames of doors and windows, decking, floor and roof tiles. MTCs offer the advantage of utilizing smaller and lower grade materials like plantation timbers to obtain larger dimension lumber for load-bearing applications [38, 39]. Cross-Laminated wood (CLT), Glued Laminated Timber (Glulam), Nail Laminated Timber (NLT), Dowel Laminated Timber (DLT), and Structural Composite Lumber (SCL) are the five most common forms of mass wood products used in building construction and are covered in this overview. Laminating allows to grade the material and control the arrangement of the lamellas to achieve the desired strength properties [60]. Because the mechanical and physical qualities of these laminates are more homogeneous than those of solid-sawn material, engineering designs can accommodate greater permitted design attributes.

Abed et al. [1] reviewed the different mass timbers available to the public thereby, encouraging the benefits of EWPs compared to traditional building materials. Layers of timber boards are stacked at right angles to one another to create CLT panels, which are then joined with a structural adhesive [53]. The wood procured for manufacturing panels is seasoned in kiln to a moisture level of 12% [34]. CLT panels of various sizes are manufactured, particularly CLTs of 18 m long by 5 m broad and 500 mm thick, which are ideal for floors, walls, and roofs. For construction applications, CLT is a competitive substitute for steel and concrete due to its many engineering advantages [53]. The technique of alternating across layers of lumber allows the end product a high level of dimensional stability, enabling the prefabrication of huge wall and floor pieces [115]. In addition, CLT has superior strength and stiffness, as well as two-way span capabilities that are comparable to those of a reinforced concrete slab, thanks to the cross-lamination process and the thicker panels. Glulam is composed of several layers of dimensional lumber with parallel grains running along the length of the member. A long-lasting, moisture-resistant glue is used to attach individual pieces of timber that are evaluated for strength based on their performance criteria. The strength-reducing qualities of every timber part are reduced in Glulam, resulting in a stronger and more dependable product than the standard solid lumber [88].

Nail Laminated Timber (NLT) is a category of engineered timber product that was initially employed over a century ago in buildings and is seeing a revival as a part of the present trend towards sustainable materials. NLT has historically been used to build factories and warehouses, but because of its appealing and advantageous performance characteristics, it also presents opportunities for a range of contemporary architectural applications. NLT has a significant advantage over other forms of mass wood in that neither it requires a specific plant nor any specialist equipment to produce. Locally accessible wood species and minimal carpentry skills are sufficient to manufacture NLT systems. Dowel Laminated Timber (DLT) is a lesser-known mass wood product initially popular in Europe but presently it is gaining popularity in North America and other nations [108]. Wooden dowels are used to join the timber parts in DLT instead of metal nails or screws as in NLT [108]. DLT panels are made by placing numerous boards of softwood planks on the edge and fitting them together by applying friction with hardwood dowels. The drier hardwood dowels expand into the surrounding lumber once they are inserted into the timber lamellas to reach moisture equilibrium. This creates a tight friction fit that enhances the dimensional stability of the panel.

Adhesives used in mass wood can be harmful to the environment because they result in formaldehyde emission and release hazardous volatile organic compounds (VOCs). As a result, the adhesives less DLT products can contribute to a better home environment by improving air quality and lowering the susceptibility of allergy responses. DLT also has the potential to replace other mass timber products using adhesives and metal fasteners because of its better recyclability and reusability characteristics [108].

Wood-polymer composites

As people's awareness of sustainability has grown, so too has interest in the creation of composite materials, such as wood plastic composites (WPC), increased. Thanks to its natural wood qualities mixed with the ability of polymers to withstand severe outdoor circumstances, it has been commercialized with significant impact on the furniture market, building, and automotive industries, among other areas. Typically, waste lignocellulosic materials such as sawdust, leftover bamboo fibers, rice husk, and other cellulose-based materials are combined with synthetic thermoplastics, such as polyvinyl chloride (PVC), high-density polyethylene (HDPE), and polypropylene (PP), to create WPCs.

Polymers are high molecular weight materials that exhibit both qualities like liquid and solid. It has flow over time that is dependent on temperature, stress, and pressure, and at the same time qualities, such as elasticity and dimensional stability. To facilitate good adhesion and processability of these composites, coupling agents, lubricants, and stabilizers are often included [102]. Certain polymers have a propensity to sag over time due to their viscoelasticity. Thermoplastics offer improved mechanical qualities, high impact resistance, and chemical resistance. These polymers are easily handled by blow moulding, extrusion, and injection moulding. WPCs, on the other hand, constitute the wood component's continuous polymer matrix surround. When woods are coupled with these matrix polymers, which are usually inexpensive commodities that flow readily when heated, significant processing flexibility is possible.

According to Schwarzkopt and Burnard [103], the primary benefits of using WPC include a decrease in reliance on non-renewable fossil-based polymers with comparable properties, a reduction in non-biodegradable plastic waste, a drop in CO2 emissions, and enhanced environmental and human health. An additional way to think of WPCs is as a product that might be used to recycle waste lignocellulosic material and/or post-consumer plastics [83]. Despite issues with high viscosities and inadequate fiber wetting, thermoplastics like HDPE and PP provide lower production cycle times and the potential to be reprocessed. Due to their low density, cheap cost, recyclable nature, biodegradability, capacity to lessen equipment abrasion, non-toxicity, and renewable nature, natural fibers have several benefits over inorganic fillers [24]. Mohammed et al. [70] in their study revealed that WPC industries employ a variety of natural fibers, including flex, hemp, and wheat. Several less useful fibers, such as areca, pineapple, banana and others, are also accessible in millions of tonnes annually but have limited value now.

Other factors affecting the preparation of natural fiber-based polymer composites are matrix selection, interface strength, processing technique, porosity, or void of the fibers. Khoaele et al. [58] provided a detailed review of the importance of process parameters on the different properties of WPCs. The authors further reviewed that the properties of the composites can be enhanced by incorporating different pretreatments and chemical treatments. The methods used for processing thermoplastic polymers for composite preparation also play a decisive step in the quality of the materials. A few of the most common methods used are hand layup followed by thermal pressing, compression moulding, extrusion and injection moulding.

Besides the automotive sector, WPCs have applications in building, construction, and so forth for laminated flooring, office items, frames, decks, and ceilings. Because WPCs are susceptible to environmental outbreaks, their applications in civil engineering are concentrated on non-load property interior components [70]. Particularly in the US, wood fiber/PE and fiber/PP have been utilized in decking projects in the building sector recently. The use of WPCs in wall insulation, floor lamination, and window and door frames in non-structural construction has grown in popularity [93]. In comparison to hardwood laminates, WPCs were shown to have better mechanical qualities for insulating structural boards and other common items.

Bamboo boards

One of the renewable resources with the fastest rate of growth is bamboo. Although globally there are more than 1200 species of bamboo, only 148 species from 29 genera may be found in India, which contributes to 30% of the world's total bamboo resource [16, 17]. China being the leader of bamboo resources, contributes to about 34% of the world's production. Bamboo is a sustainable lignocellulosic material and as a structural material in composite form after processing has gained worldwide acceptance, the end product is known as Structural Bamboo Composite (SBC). The general fabrication of SBCs involves cutting long bamboo culms into desirable shorter lengths followed by knot removal, splitting and removal of outer skin. Subsequently, softening and defibering are done to prepare the raw materials. The final stage involves the adhesive application and the pressing parameters involved. The processing technique applied to prepare bamboo composite manufacturing has a considerable impact on its performance [54]. A compiled list of the effect of bamboo species, adhesive type and spread rate, and the effect of pressing parameters on mechanical and physical properties of SBCs has been given by Kelkar et al. [56].

Particleboard [43], plywood, laminated composites, oriented strand board, zephyr board, cement-bonded board, and wood polymer composites can be derived from bamboo-based composites after slight value addition, thanks to the recent advancements in manufacturing technology [20, 55, 56]. Bamboo composites, which are created by joining various grades of bamboo culms, have been highlighted in some recent studies. Bahtiar et al. [8] revealed this process also removes the unpredictable nature of bamboo and provides consistent serviceability and safety. Because bamboo composites have so many advantages over wood and wood-based products—including superior performance in load-bearing applications, ecological benefits, ease of availability, and low carbon footprints—they can be viewed as a competitive option in structural applications [42, 116]. When utilized in dry conditions, laminated bamboo timber has remarkable properties as a structural material, including high strength, stiffness, minimal deformation, and dimensional stability. It can be effectively used as building materials as shown in Fig. 2.

Fig. 2 Hut made of bamboo laminates at Forest Research Institute, Dehradun [98], open access

To obtain specified forms and sizes, the SBCs generally utilize processed bamboo in the form of strips, strands, or flattened panels [65] that have been joined with structural-grade polymeric adhesives and then compressed [106]. Other than SBCs, bamboo composites like Laminated bamboo lumber (LBL), bamboo scrimber (SBL), and bamboo woven panels have become popular in the field of structural usage. Bamboo laminated boards, veneers and panels are the modern choice of building materials. Additionally, bamboo can be utilized to create aesthetically pleasing and useful door and window frames [98]. Japan and some western countries are popularly using bamboo panel flooring due to its soft-natural appearance and long-lasting gloss [31]. Bamboo composites not only provide sustainable materials with consistent cross sections [18, 19], but they also have exceptional mechanical properties that rarely vary [120]. Bamboo panels offer superior sound absorption because of their capacity to absorb sound waves [101]. Degradation against dry wood and subterranean termites could be achieved by treating bamboo composites with preservative treatments like permethrin, deltamethrin, and combination of the two [5]. Treated bamboo composites with mono ammonium phosphate combined with boric acid, incorporated with phenol formaldehyde resins have found to have satisfactory fire retardancy [51, 67]. An optimized combination of acrylic emulsion, melamine, ammonium polyphosphate, pentaerythritol, and titanium dioxide is an alternate effective fire-retardant chemical for composite bamboo lumbers [66] and nanoparticles of calcium carbonate, calcium oxide, magnesium carbonate, and magnesium oxide show effective fire performance [22, 28, 61].

The development of high-strength bamboo-based composites would reduce the demand and supply gap of the wood and would change the perspective towards other lignocellulosic materials. The [46] INBAR studies positioned the global trade value of bamboo panels at 135 million USD, with the United States, the European Union, and Singapore as the biggest importers. Technological involvements have improvised the overall performance of the composites and made it at par with the conventional timbers being used. However, since there is still a problem with its emissions, using adhesives that contain formaldehyde continues to be one of the primary restrictions. The development of standard preparative codes and stringent testing guidelines would ensure homogeneity and quality products.

5 Conclusion

Engineered wood products are contemporary materials that have several advantages over typical solid wood items. In certain situations, they are more durable, stronger, and more consistent than conventional products, and they can be created from reclaimed wood waste. This reduces waste and conserves resources while also making them a more sustainable and environmentally friendly alternative to traditional wood goods. A summary of the most popular synthetic adhesives is discussed in this study, with a focus on their advantages for the wood industry. Synthetic adhesives are not new, and they perform well, but using synthetic adhesives generated from unsustainable sources has considerable negative impacts on the environment and human health, which has raised concerns about the usage of these adhesives. As an alternate technique, researchers have focused on the development of sustainable bio-based wood adhesives derived from lignocellulosic biomass wastes since they have proven to be a cost-effective, environmentally friendly, and health-promoting alternative. Regarding sustainable wood adhesives made from naturally occurring renewable feedstocks, the current study offers extensive information like adhesives from lignin, tannin, and protein. Engineered woods are very flexible, with applications ranging from building to furniture and packaging. There are several non-wood fibrous materials that can be utilized to produce boards. These would not only reduce loads of waste but could also help in carbon sequestration. However, to embrace new developments in industries and promote the commercialization of sustainable wood composites, numerous technological obstacles must be overcome. The newer technologies discussed in this review, like, enzymatically modified boards and fungal-based boards can open arenas towards sustainable production techniques. These areas need renewed research interest so that these technologies can be transferred to the industries. These are getting more popular as the need for environmentally friendly building materials develops. Wood is a natural carbon sink, which means it absorbs CO_2 from the atmosphere. Carbon is kept in the product when wood is utilized to make these composites. As the world is becoming more and more aware of sustainability and the need to protect the environment, engineered composites are expected to become even more popular.

Acknowledgements The authors would like to thank the Director and Group Coordinator Research of the Institute of Wood Science and Technology, Bangalore, India, for extending support and motivation for carrying out the work. The authors would also like to acknowledge the support of Dr. Rakesh Kumar, Scientist—F at the Institute of Wood Science and Technology, Bangalore, India, for his motivation.

References

1. Abed, J., Rayburg, S., Rodwell, J., Neave, M.: A review of the performance and benefits of mass timber as an alternative to concrete and steel for improving the sustainability of structures. Sustainability **14**(9), 5570 (2022)
2. Acda, M.N., Cabangon, R.J.: Termite resistance and physico-mechanical properties of particleboard using waste tobacco stalk and wood particles. Int. Biodeterior. Biodegrad. **85**, 354–358 (2013)
3. Ahankari, S.S., Mohanty, A.K., Misra, M.: Mechanical behaviour of agro-residue reinforced poly (3-hydroxybutyrate-co-3-hydroxyvalerate), (PHBV) green composites: a comparison with traditional polypropylene composites. Compos. Sci. Technol. **71**(5), 653–657 (2011)
4. Aladejana, J.T., Wu, Z., Fan, M., Xie, Y.: Key advances in development of straw fibre biocomposite boards: an overview. Mater. Res. Express **7**(1), 012005 (2020)
5. Alipon, M.A., Garcia, C.M., Bondad, E.O.: Glue and preservative effects on the properties and durability of engineered bamboo boards. Philipp. J. Sci. **147**(4), 601–616 (2018)
6. Alwani, M.S., Khalil, H.A., Asniza, M., Suhaily, S.S., Amiranajwa, A.N., Jawaid, M.: Agricultural biomass raw materials: the current state and future potentialities. Biomass Bioenergy: Process. Prop. 77–100 (2014)
7. Andrade, P.I., Araujo, S.D.O., Neiva, D.M., Vital, B.R., Oliveira Carneiro, A.D.C., Gominho, J., Pereira, H.: Strength properties and dimensional stability of particleboards with different proportions of thermally treated recycled pine particles. Holzforschung **70**(5), 467–474 (2016)
8. Bahtiar, E.T., Denih, A., Putra, G.R.: Multi-culm bamboo composites as sustainable materials for green constructions: section properties and column behavior. Results Eng. **17**, 100911 (2023)
9. Berthet, M.A., Angellier-Coussy, H., Machado, D., Hilliou, L., Staebler, A., Vicente, A., Gontard, N.: Exploring the potentialities of using lignocellulosic fibres derived from three food by-products as constituents of biocomposites for food packaging. Ind. Crops Prod. **69**, 110–122 (2015)
10. Brischke, C., Stricker, S., Meyer-Veltrup, L., Emmerich, L.: Changes in sorption and electrical properties of wood caused by fungal decay. Holzforschung **73**(5), 445–455 (2019)
11. Bruscato, C., Malvessi, E., Brandalise, R.N., Camassola, M.: High performance of macrofungi in the production of mycelium-based biofoams using sawdust—sustainable technology for waste reduction. J. Clean. Prod. **234**, 225–232 (2019)
12. Buzo, A.L.S.C., Silva, S.A.M., De Moura Aquino, V.B., Chahud, E., Branco, L.A.M.N., De Almeida, D.H., Christoforo, A.L., Almeida, J.P., Lahr, F.A.R.: Addition of sugarcane bagasse for the production of particleboards bonded with urea-formaldehyde and polyurethane resins. Wood Res. **65**, 727–736 (2020)
13. Cai, Z.: Selected properties of MDF and flakeboard overlaid with fiberglass mats. For. Prod. J. **56**(11/12), 142–146 (2006)
14. Cai, Z., Ross, R.J.: Mechanical properties of wood-based composite materials. In: Wood Handbook-Wood as an Engineering Material. Centennial Edition. General Technical Report FPL-GTR-190. Department of Agriculture, Forest Service, Forest Products Laboratory. Madison, Wisconsin, EUA, 12 (2010)

15. Chandel, A.K., da Silva, S.S., Singh, O.V.: Detoxification of lignocellulosic hydrolysates for improved bioethanol production. Biofuel Prod.-Recent Dev. Prospect. **10**, 225 (2011)
16. Chaowana, P.: Bamboo: an alternative raw material for wood and wood-based composites. J. Mater. Sci. Res. **2**(2), 90 (2013)
17. Chawla, V.K., Seshasayee, V., Yadav, R.: Use of lignocellulosic biomaterials for sustainable development of bamboo strand lumber for structural applications. BioResources **17**(2), 2864 (2022)
18. Chen, G., Yu, Y., Li, X., He, B.: Mechanical behavior of laminated bamboo lumber for structural application: an experimental investigation. Eur. J. Wood Wood Prod. **78**(1), 53–63 (2020)
19. Chen, M., Ye, L., Li, H., Wang, G., Chen, Q., Fang, C., Dai, C., Fei, B.: Flexural strength and ductility of moso bamboo. Constr. Build. Mater. **246**, 118418 (2020)
20. Dauletbek, A., Li, H., Lorenzo, R., Corbi, I., Corbi, O., Ashraf, M.: A review of basic mechanical behavior of laminated bamboo lumber. J. Renew. Mater. **10**(2), 273 (2022)
21. Drzal, L.T., Mohanty, A.K., Burgueno, R., Misra, M.: Biobased structural composite materials for housing and infrastructure applications: opportunities and challenges. In: Proceedings of the NSF Housing Research Agenda Workshop, vol. 2, pp. 129–140 (2004, February)
22. Du, C.G., Song, J.G., Chen, Y.X.: The effect of applying methods of fire retardant on physical and mechanical properties of bamboo scrimber. Adv. Mater. Res. **1048**, 465–468 (2014)
23. Elsacker, E., Vandelook, S., Van Wylick, A., Ruytinx, J., De Laet, L., Peeters, E.: A comprehensive framework for the production of mycelium-based lignocellulosic composites. Sci. Total. Environ. **725**, 138431 (2020)
24. Facca, A.G., Kortschot, M.T., Yan, N.: Predicting the elastic modulus of natural fiber reinforced thermoplastics. Compos. A Appl. Sci. Manuf. **37**(10), 1660–1671 (2006)
25. Felby, C., Hassingboe, J., Lund, M.: Pilot-scale production of fiberboards made by laccase oxidized wood fibers: board properties and evidence for cross-linking of lignin. Enzyme Microb. Technol. **31**(6), 736–741 (2002)
26. Felby, C., Nielsen, B.R., Olesen, P.O., Skibsted, L.H.: Identification and quantification of radical reaction intermediates by electron spin resonance spectrometry of laccase-catalyzed oxidation of wood fibers from beech (*Fagus sylvatica*). Appl. Microbiol. Biotechnol. **48**(4), 459–464 (1997)
27. Frihart, C.R.: Wood adhesion and adhesives. In: Handbook of Wood Chemistry and Wood Composites, pp. 215–278 (2005)
28. Fu, B., Li, X., Yuan, G., Chen, W., Pan, Y.: Preparation and flame retardant and smoke suppression properties of bamboo-wood hybrid scrimber filled with calcium and magnesium nanoparticles. J. Nanomater. **2014**, 3–3 (2014)
29. Gertjejansen, R.O., Haygreen, J.G., French, D.W.: Particleboard from aspen flakes and sunflower hulls (1972)
30. Ghofrani, M., Haghdan, S., NicKhah, V., Ahmadi, K.: Improvement of physical and mechanical properties of particleboard made of apple tree pruning and sunflower stalk using titanium oxide nanoparticles. Eur. J. Wood Wood Prod. **73**, 661–666 (2015)
31. Gupta, A., Kumar, A.: Potential of bamboo in sustainable development. Asia Pac. Bus. Rev. **4**(3), 100–107 (2008)
32. Han, G., Zhang, C., Zhang, D., Umemura, K., Kawai, S.: Upgrading of urea formaldehyde-bonded reed and wheat straw particleboards using silane coupling agents. J. Wood Sci. **44**, 282–286 (1998)
33. Hannah Mason: Natural fiber composites: growing to fit sustainability needs. Composites World (2023). https://www.compositesworld.com/articles/natural-fiber-composites-growing-to-fit-sustainability-needs. Accessed 27 Mar 2023
34. Harte, A.M.: Mass timber–the emergence of a modern construction material. J. Struct. Integr. Maint. **2**(3), 121–132 (2017)
35. Heinze, T., Petzold, K.: Cellulose chemistry: novel products and synthesis paths. In: Monomers, Polymers and Composites from Renewable Resources. Elsevier, pp. 343–368 (2008)

36. Hejna, A., Formela, K., Saeb, M.R.: Processing, mechanical and thermal behavior assessments of polycaprolactone/agricultural wastes biocomposites. Ind. Crops Prod. **76**, 725–733 (2015)
37. Hellmayr, R., Bischof, S., Wühl, J., Guebitz, G.M., Nyanhongo, G.S., Schwaiger, N., Liebner, F., Wimmer, R.: Enzymatic conversion of lignosulfonate into wood adhesives: a next step towards fully biobased composite materials. Polymers **14**(2), 259 (2022)
38. Herawati, E., Massijaya, M.Y., Nugroho, N.: Performance of glued laminated beams made from small diameter fast-growing tree species. J. Biol. Sci. **10**(1), 37–42 (2010)
39. Hermawan, D., Hadi, Y.S., Fajriani, E., Massijaya, M.Y., Hadjib, N.: Resistance of particleboards made from fast-growing wood sspecies to subterranean termite attack. Insects **3**(2), 532–537 (2012)
40. Hidayat, W., Aprilliana, N., Asmara, S., Bakri, S., Hidayati, S., Banuwa, I.S., Lubis, M.A.R., Iswanto, A.H.: Performance of eco-friendly particleboard from agro-industrial residues bonded with formaldehyde-free natural rubber latex adhesive for interior applications. Polym. Compos. **43**(4), 2222–2233 (2022)
41. Hong, N.V., Pyka, G., Wevers, M., Goderis, B., Van Puyvelde, P., Verpoest, I., Van Vuure, A.W.: Processing rigid wheat gluten biocomposites for high mechanical performance. Compos. A Appl. Sci. Manuf. **79**, 74–81 (2015)
42. Huang, D., Bian, Y., Zhou, A., Sheng, B.: Experimental study on stress–strain relationships and failure mechanisms of parallel strand bamboo made from phyllostachys. Constr. Build. Mater. **77**, 130–138 (2015)
43. Huang, Y., Ji, Y., Yu, W.: Development of bamboo scrimber: A literature review. J. Wood Sci. **65**(1), 1–10 (2019)
44. Huaxu, Z., Hua, L.S., Tahir, P.M., Ashaari, Z., Al-Edrus, S.S.O., Ibrahim, N.A., Abdullah, L.C., Mohamad, S.F.: Physico-mechanical and biological durability of citric acid-bonded rubberwood particleboard. Polymers **13**(1), 98 (2020)
45. Hussin, M.H., Abd Latif, N.H., Hamidon, T.S., Idris, N.N., Hashim, R., Appaturi, J.N., Brosse, N., Ziegler-Devin, I., Chrusiel, L., Fatriasari, W., Syamani, F.A., Iswanto, A.H., Hua, L.S., Al Edrus, S.S.A.O., Lum, W.C., Antov, P., Savov, V., Lubis, M.A.R., Kristak, L., Reh, R., Sedliačik, J.: Latest advancements in high-performance bio-based wood adhesives: a critical review. J. Mater. Res. Technol. **21**, 3909–3946 (2022)
46. INBAR: In: Organisation, I.B.A.R. (ed.) Trade Overview 2018 Bamboo and Rattan Commodities in China. INBAR, China (2019)
47. Islam, M.R., Tudryn, G., Bucinell, R., Schadler, L., Picu, R.C.: Mechanical behavior of mycelium-based particulate composites. J. Mater. Sci. **53**, 16371–16382 (2018)
48. Islam, M.R., Tudryn, G., Bucinell, R., Schadler, L., Picu, R.C.: Morphology and mechanics of fungal mycelium. Sci. Rep. **7**(1), 13070 (2017)
49. Janiszewska, D., Frąckowiak, I., Mytko, K.: Exploitation of liquefied wood waste for binding recycled wood particleboards. Holzforschung **70**(12), 1135–1138 (2016)
50. Jawaid, M.H.P.S., Khalil, H.A.: Cellulosic/synthetic fibre reinforced polymer hybrid composites: a review. Carbohyd. Polym. **86**(1), 1–18 (2011)
51. Jin, X.B., Jiang, Z.H., Wen, X.W., Zhang, R., Qin, D.C.: Flame retardant properties of laminated bamboo lumber treated with monoammonium phosphate (MAP) and boric acid/borax (SBX) compounds. BioResources **12**(3), 5071–5085 (2017)
52. Jones, M., Mautner, A., Luenco, S., Bismarck, A., John, S.: Engineered mycelium composite construction materials from fungal biorefineries: a critical review. Mater. Des. **187**, 108397 (2020)
53. Karacabeyli, E., Douglas, B.: CLT handbook. In: US Edition, FPInnovations and Binational Softwood Lumber Council, Point-Claire, Quebec (2013)
54. Kelkar, B.U., Sharma, S.K., Shukla, S.R.: Comparative performance of phenol formaldehyde-bonded laminated bamboo lumber and bamboo strand lumber prepared from four different bamboo species. J. Trop. For. Sci. **33**(4), 482–493 (2021)
55. Kelkar, B.U., Shukla, S.R., Nagraik, P., Paul, B.N.: Structural bamboo composites: a review of processing, factors affecting properties and recent advances. Adv. Bamboo Sci. **3**, 100026 (2023)

56. Kelkar, B.U., Shukla, S.R., Yadav, S.M., Bansal, R.: Performance of laminated bamboo lumber and bamboo strand lumber coated with solvent and water-based polyurethane against accelerated UV and natural weathering. Ind. Crops Prod. **192**, 116058 (2023)
57. Khazaeian, A., Ashori, A., Dizaj, M.Y.: Suitability of sorghum stalk fibers for production of particleboard. Carbohyd. Polym. **120**, 15–21 (2015)
58. Khoaele, K.K., Gbadeyan, O.J., Chunilall, V., Sithole, B.: A review on waste wood reinforced polymer composites and their processing for construction materials. Int. J. Sustain. Eng. **16**(1), 104–116 (2023)
59. Klímek, P., Meinlschmidt, P., Wimmer, R., Plinke, B., Schirp, A.: Using sunflower (Helianthus annuus L.), topinambour (Helianthus tuberosus L.) and cup-plant (Silphium perfoliatum L.) stalks as alternative raw materials for particleboards. Ind. Crops Prod. **92**, 157–164 (2016)
60. Komariah, R.N., Hadi, Y.S., Massijaya, M.Y., Suryana, J.: Physical mechanical properties of glued laminated timber made from tropical small-diameter logs grown in Indonesia. J. Korean Wood Sci. Technol. **43**(2), 156–167 (2015)
61. Kushwaha, P.K., Chawla, V., Nandanwar, A., Mamtha, B.S., Murhty, N.: Study on the effect of nanoparticles on fire resistance and smoke suppression properties of bamboo strand lumber. J. Indian Acad. Wood Sci. **18**(1), 32–38 (2021)
62. Lee, S.H., Ashaari, Z., Lum, W.C., Ang, A.F., Abdul Halip, J., Halis, R.: Chemical, physico-mechanical properties and biological durability of rubberwood particleboards after post heat-treatment in palm oil. Holzforschung **72**(2), 159–167 (2018)
63. Lee, S.H., Lum, W.C., Boon, J.G., Kristak, L., Antov, P., Pędzik, M., Rogoziński, T., Taghiyari, H.R., Lubis, M.A.R., Fatriasari, W., Yadav, S.M., Chotikhun, A., Pizzi, A.: Particleboard from agricultural biomass and recycled wood waste: a review. J. Mater. Res. Technol. **20**, 4630–4658 (2022)
64. Li, X., Cai, Z., Winandy, J.E., Basta, A.H.: Selected properties of particleboard panels manufactured from rice straws of different geometries. Bioresour. Technol. **101**(12), 4662–4666 (2010)
65. Liu, X., Smith, G.D., Jiang, Z., Bock, M.C., Boeck, F., Frith, O., Gatóo, A., Liu, K., Mulligan, H., Semple, K.E., Sharma, B., Ramage, M.: Nomenclature for engineered bamboo (2016)
66. Lv, Q., Wang, W., Liu, Y.: Charring depth and charring rate of cross-laminated bamboo slabs exposed to a one-sided standard fire. Fire Saf. J. **125**, 103439 (2021)
67. Madden, J., Guiterrez, M., Maluk, C.: Structural performance of laminated bamboo columns during fire. In: Australian Structural Engineering Conference: ASEC-2018. Engineers Australia, vol. 210, January 2018 (2018)
68. Mayer, A.M., Staples, R.C.: Laccase: new functions for an old enzyme. Phytochemistry **60**(6), 551–565 (2002)
69. Mendes, R.F., Mendes, L.M., Oliveira, S.L., Freire, T.P.: Use of sugarcane bagasse for particleboard production. Key Eng. Mater. **634**, 163–171 (2015)
70. Mohammed, L., Ansari, M.N., Pua, G., Jawaid, M., Islam, M.S.: A review on natural fiber reinforced polymer composite and its applications. Int. J. Polym. Sci. **2015**, 243947 (2015)
71. Mohanty, A., Vivekanandhan, S., Pin, J., Misra, M.: Composites from renewable and sustainable resources: challenges and innovations. Science **362**, 536–542 (2018). https://doi.org/10.1126/science.aat9072
72. Moor, H., Eggers, J., Fabritius, H., Forsell, N., Henckel, L., Bradter, U., Mazziotta, A., Nordén, J., Snäll, T.: Rebuilding green infrastructure in boreal production forest given future global wood demand. J. Appl. Ecol. **59**(6), 1659–1669 (2022)
73. Müller, C., Schwarz, U., Thole, V.: Zur Nutzung von Agrar-Reststoffen in der Holzwerkstoffindustrie. Eur. J. Wood Wood Prod. **70**(5), 587–594 (2012)
74. Nasir, M., Gupta, A., Beg, M.D.H., Chua, G.K., Asim, M.: Laccase application in medium density fiberboard to prepare a bio-composite. RSC Adv. **4**(22), 11520–11527 (2014)
75. Nasir, M., Gupta, A., Beg, M.D.H., Chua, G.K., Kumar, A.: Fabrication of medium density fiberboard from enzyme treated rubber wood (Hevea brasiliensis) fiber and modified organosolv lignin. Int. J. Adhes. Adhes. **44**, 99–104 (2013)

76. Nasir, M., Gupta, A., Beg, M.D.H., Chua, G.K., Jawaid, M., Kumar, A., Khan, T.A.: Fabricating eco-friendly binderless fiberboard from laccase-treated rubber wood fiber. BioResources **8**(3), 3599–3608 (2013)
77. Nasir, M., Hashim, R., Sulaiman, O., Gupta, A., Khan, T.A., Jawaid, M., Asim, M.: Natural fiber improvement by laccase; optimization, characterization and application in medium density fiberboard. J. Nat. Fibers **14**(3), 379–389 (2017)
78. Nasir, M., Hashim, R., Sulaiman, O., Nordin, N.A., Lamaming, J., Asim, M.: Laccase, an emerging tool to fabricate green composites: a review. BioResources **10**(3), 6262–6284 (2015)
79. Nasir, M., Sulaiman, O., Hashim, R., Hossain, K., Gupta, A., Asim, M.: Rubberwood fiber treatment by laccase enzyme and its application in medium density fiberboard. J Pure Appl Microbiol **9**(3), 2095–2100 (2015)
80. Nazerian, M., Beyki, Z., Gargarii, R.M., Kool, F.: The effect of some technological production variables on mechanical and physical properties of particleboard manufactured from cotton (Gossypium hirsutum) stalks. Maderas. Ciencia y tecnología **18**(1), 167–178 (2016)
81. Neitzel, N., Hosseinpourpia, R., Walther, T., Adamopoulos, S.: Alternative materials from agro-industry for wood panel manufacturing—a review. Materials **15**(13), 4542 (2022)
82. Nemli, G., Demirel, S., Gümüşkaya, E., Aslan, M., Acar, C.: Feasibility of incorporating waste grass clippings (Lolium perenne L.) in particleboard composites. Waste Manag. **29**(3), 1129–1131 (2009)
83. Niska, K.O., Sain, M. (eds.): Wood-Polymer Composites. Elsevier (2008)
84. Nkeuwa, W.N., Zhang, J., Semple, K.E., Chen, M., Xia, Y., Dai, C.: Bamboo-based composites: a review on fundamentals and processes of bamboo bonding. Compos. B Eng. **235**, 109776 (2022)
85. Nyanhongo, G.S., Nugroho Prasetyo, E., Herrero Acero, E., Guebitz, G.M.: Engineering strategies for successful development of functional polymers using oxidative enzymes. Chem. Eng. Technol. **35**(8), 1359–1372 (2012)
86. Oh, Y.S., Jamaludin, M.A.: Evaluation of rapeseed stalk particleboard bonded with laboratory-made urea-formaldehyde resin. Ciência Florestal **25**, 515–521 (2015)
87. Oh, Y.S., Yoo, J.Y.: Properties of particleboard made from chili pepper stalks. J. Trop. For. Sci. 473–477 (2011)
88. Ong, C.B.: Glue-laminated timber (Glulam). In: Wood Composites, pp. 123–140 (2015)
89. Ortner, A., Huber, D., Haske-Cornelius, O., Weber, H.K., Hofer, K., Bauer, W., Guebitz, G.M.: Laccase mediated oxidation of industrial lignins: Is oxygen limiting? Process Biochem. **50**(8), 1277–1283 (2015)
90. Panthapulakkal, S., Sain, M.: Agro-residue reinforced high-density polyethylene composites: fiber characterization and analysis of composite properties. Compos. A Appl. Sci. Manuf. **38**(6), 1445–1454 (2007)
91. Paul, W., Ohlmeyer, M., Leithoff, H., Boonstra, M.J., Pizzi, A.: Optimising the properties of OSB by a one-step heat pre-treatment process. Holz als Roh-und Werkstoff **64**(3), 227–234 (2006)
92. Pędzik, M., Janiszewska, D., Rogoziński, T.: Alternative lignocellulosic raw materials in particleboard production: a review. Ind. Crops Prod. **174**, 114162 (2021)
93. Pickering, K.L., Efendy, M.A., Le, T.M.: A review of recent developments in natural fiber composites and their mechanical performance. Compos. A Appl. Sci. Manuf. **83**, 98–112 (2016)
94. Pirayesh, H., Khanjanzadeh, H., Salari, A.: Effect of using walnut/almond shells on the physical, mechanical properties and formaldehyde emission of particleboard. Compos. B Eng. **45**(1), 858–863 (2013)
95. Rana, A.K., Mandal, A., Bandyopadhyay, S.: Short jute fiber reinforced polypropylene composites: effect of compatibiliser, impact modifier and fiber loading. Compos. Sci. Technol. **63**(6), 801–806 (2003)
96. Rao, P.S.: Chemical constituents of wood. In: Handbook on Indian Wood and Wood Panels. Oxford University Press, pp. 207–250 (1988)

97. Ray, S., Kumar, R.: A snapshot review on enzymatic modification of wood. MRS Adv. **7**(27), 562–569 (2022)
98. Ray, S., Kumar, R., Sharma, R.: Bamboo: the green gold. Planta **4**, 941–951 (2022)
99. Rowell, R.M.: The state of art and future development of bio-based composite science and technology towards the 21st century. In: Proceedings, 4th Pacific Rim Bio-Based Composite Symposium, vol. 1 (1998)
100. Sakaguchi, H., Uyama, N., Uyama, H.: Preserving boiled eggs with a sterilization system employing microbial laccase and wood vinegar. Anim. Sci. J. **78**(6), 668–671 (2007)
101. Sawarkar, A.D., Shrimankar, D.D., Kumar, A., Kumar, A., Singh, E., Singh, L., Kumar, R.: Commercial clustering of sustainable bamboo species in India. Ind. Crops Prod. **154**, 112693 (2020)
102. Schneider, M.H.: Wood polymer composites. In: Wood and Fiber Science, pp. 142–151 (1994)
103. Schwarzkopf, M.J., Burnard, M.D.: Wood-plastic composites—performance and environmental impacts. In: Environmental Impacts of Traditional and Innovative Forest-Based Bioproducts, pp. 19–43 (2016)
104. Semple, K.E., Vnučec, D., Kutnar, A., Kamke, F.A., Mikuljan, M., Smith, G.D.: Bonding of THM modified Moso bamboo (Phyllostachys pubescens Mazel) using modified soybean protein isolate (SPI) based adhesives. Eur. J. Wood Wood Prod. **73**, 781–792 (2015)
105. Shang, L., Ahrenfeldt, J., Holm, J.K., Sanadi, A.R., Barsberg, S., Thomsen, T., Stelte, W., Henriksen, U.B.: Changes of chemical and mechanical behavior of torrefied wheat straw. Biomass Bioenergy **40**, 63–70 (2012)
106. Sharma, B., van der Vegte, A.: Engineered bamboo for structural applications. In: Nonconventional and Vernacular Construction Materials. Woodhead Publishing, pp. 597–623 (2020)
107. Smole, M.S., Kleinschek, K.S., Kreze, T., Strnad, S., Mandl, M., Wachter, B.: Physical properties of grass fibres. Chem. Biochem. Eng. Q. **18**(1), 47–54 (2004)
108. Sotayo, A., Bradley, D., Bather, M., Sareh, P., Oudjene, M., El-Houjeyri, I., Harte, A.M., Mehra, S., O'Ceallaigh, C., Haller, P., Namari, S., Makradi, A., Belouettar, S., Bouhala, L., Deneufbourg, F., Guan, Z.: Review of state of the art of dowel laminated timber members and densified wood materials as sustainable engineered wood products for construction and building applications. Dev. Built Environ. **1**, 100004 (2020)
109. Spiridon, I., Popa, V.I.: Hemicelluloses: major sources, properties and applications. In: Monomers, Polymers and Composites from Renewable Resources. Elsevier, pp. 289–304 (2008)
110. Sun, W., Tajvidi, M., Howell, C., Hunt, C.G.: Functionality of surface mycelium interfaces in wood bonding. ACS Appl. Mater. Interfaces **12**(51), 57431–57440 (2020)
111. Sustainability standards provide benefits for trade in timber products, 28 September 2018. UNEP (2018). http://www.unep.org/news-and-stories/story/sustainability-standards-provide-benefits-trade-timber-products
112. Syahfitri, A., Hermawan, D., Kusumah, S.S., Ismadi, Lubis, M.A.R., Widyaningrum, B.A., Ismayati, M., Amanda, P., Ningrum, R.S., Sutiawan, J.: Conversion of agro-industrial wastes of sorghum bagasse and molasses into lightweight roof tile composite. Biomass Convers. Biorefin. **14**, 1–15 (2022)
113. Taha, I., Elkafafy, M.S., El Mously, H.: Potential of utilizing tomato stalk as raw material for particleboards. Ain Shams Eng. J. **9**(4), 1457–1464 (2018)
114. UNCTAD: Global trade slows, but 'green goods' grow, March 2013 (2013). https://unctad.org/news/global-trade-slows-green-goods-grow
115. Van De Kuilen, J.W.G., Ceccotti, A., Xia, Z., He, M.: Very tall wooden buildings with cross laminated timber. Procedia Eng. **14**, 1621–1628 (2011)
116. Van der Lugt, P., Van den Dobbelsteen, A.A.J.F., Janssen, J.J.A.: An environmental, economic and practical assessment of bamboo as a building material for supporting structures. Constr. Build. Mater. **20**(9), 648–656 (2006)
117. Wang, W., Zhu, Y., Cao, J., Guo, X.: Thermal modification of Southern pine combined with wax emulsion pre impregnation: effect on hydrophobicity and dimensional stability. Holzforschung **69**(4), 405–413 (2015)

118. Widsten, P., Kandelbauer, A.: Laccase applications in the forest products industry: a review. Enzyme Microb. Technol. **42**(4), 293–307 (2008)
119. Wood: Wood strengthens sustainability commitments, 30 March 2021. Wood (2021). https://www.woodplc.com/news/latest-press-releases/2021/wood-strengthens-sustainability-commitments. Accessed 26 Feb 2024
120. Wu, M., Mei, L., Guo, N., Ren, J., Zhang, Y., Zhao, Y.: Mechanical properties and failure mechanisms of engineering bamboo scrimber. Constr. Build. Mater. **344**, 128082 (2022)
121. Xing, W., Hao, J., Sikora, K.S.: Shear performance of adhesive bonding of cross-laminated bamboo. J. Mater. Civ. Eng. **31**(9), 04019201 (2019)
122. Xu, J., Widyorini, R., Kawai, S.: Properties of kenaf core binderless particleboard reinforced with kenaf bast fiber-woven sheets. J. Wood Sci. **51**(4), 415–420 (2005)
123. Yalinkilic, M.K., Imamura, Y., Takahashi, M., Kalaycioglu, H., Nemli, G., Demirci, Z., Ozdemir, T.: Biological, physical and mechanical properties of particleboard manufactured from waste tea leaves. Int. Biodeterior. Biodegrad. **41**(1), 75–84 (1998)
124. Yano, B.B., Silva, S.A., Almeida, D.H.D., Aquino, V.B.D.M., Christoforo, A.L., Rodrigues, E.F., Carvalho Junior, A.N., Silva, A.P., Lahr, F.A.R.: Use of sugarcane bagasse and industrial timber residue in particleboard production. BioResources **15**(3), 4753–4762 (2020)
125. Ye, X.P., Julson, J., Kuo, M., Womac, A., Myers, D.: Properties of medium density fiberboards made from renewable biomass. Bioresour. Technol. **98**(5), 1077–1084 (2007)
126. Zabel, R.A., Morrell, J.J.: Chapter Five—Fungal metabolism in relation to wood decay. In: Wood Microbiology, pp. 129–148 (2020)
127. Zhou, G., Li, J., Chen, Y., Zhao, B., Cao, Y., Duan, X., Cao, Y.: Determination of reactive oxygen species generated in laccase catalyzed oxidation of wood fibers from Chinese fir (*Cunninghamia lanceolata*) by electron spin resonance spectrometry. Bioresour. Technol. **100**(1), 505–508 (2009)
128. Zini, E., Scandola, M.: Green composites: an overview. Polym. Compos. **32**(12), 1905–1915 (2011)

Lignin-Based Wood Adhesives

Ega Cyntia Watumlawar and Byung-Dae Park

Abstract Lignin is the most abundant aromatic natural compound and has become a promising material in replacing petroleum-based polymers and introducing advanced materials to wood adhesives. This chapter provides an up-to-date overview of the lignin modifications reported in recent years to synthesize lignin-based wood adhesives. These modifications consist of answers to the challenges faced in the utilization of lignin with different treatments and conditions to prepare versatile lignin-based materials. This chapter focuses on the various aspects of lignin-based wood adhesives such as the introduction of lignin to synthetic polymers, advanced routes of chemical modification, and applications of lignin-based adhesives. The use of lignin to replace formaldehyde-based adhesives is expected to mitigate climate change and contribute to the development of bio-based adhesives for greener wood-based materials in the future.

Keywords Lignin · Lignin modification · Lignin-based thermoset · Wood adhesive

1 Introduction

Lignin is the most abundant aromatic bio-based polymer in the world, contributing 15–30% of lignocellulosic biomass [117]. Technical lignin obtained from pulping and paper-making industries includes Kraft lignin, soda lignin, organosolv lignin, sulfonate lignin, and enzymatic hydrolysis lignin [37]. However, only 5% of 70 million tons of industrial lignin is utilized while 95% of the lignin is burned to acquire energy for the industries [66]. Characteristics of these lignins have been studied for synthetic thermoplastic, polyurethane, phenolic and epoxy resin, engineering, fibers, adhesives, etc., owing to their remarkable aromatic structure and various reactive

E. C. Watumlawar · B.-D. Park (✉)
Department of Wood and Paper Science, Kyungpook National University, Daegu, Republic of Korea
e-mail: byungdae@knu.ac.kr

S. S. Osman Al Edrus et al. (eds.), *Green Lignocellulosic-Based Panels*, Engineering Materials, https://doi.org/10.1007/978-981-96-3404-0_2

functional groups. Moreover, the advantages such as high thermal stability, nontoxicity, biodegradability, renewability, and low costs make lignin potential biomass for converting into many high-value-added products in the future [1, 3, 109, 110, 113, 144].

Paper and pulp industry is crowned as one of the largest industrial commodities and regarded as an important lignin producers with 50 and 70 Mt per year [133]. The pulp industry is divided into five main processes employed to separate technical lignin such as sulfite, Kraft, organosolv, soda, and enzymatic hydrolysis process [42, 68, 111]. Some processes like lignosulfonate, Kraft, and soda lignin are produced in large quantities at commercial scales, including Lignoforce, LignoBoost, Howard methods, etc. Meanwhile, organosolv and enzymatic hydrolysis lignin lack commercial production methods because they are produced only in small laboratories.

Lignin is heterogeneous and insoluble in most organic solvents, and incompatible with other materials because of its complicated cross-linked structure [94]. The aromaticity of lignin encouraged chemical modification such as polymerization, demethylation, allylation, etc. which is the essence for various material applications and has attracted great attention in recent years. Lignin shows good prospects due to its complex polymeric structure of multi-substituted phenolic molecules derived primarily from three hydrocinnamyl alcohols or monolignols [37]. Among the various applications, thermosetting lignin polymers based on phenol-formaldehyde (PF) resins, epoxy resins, and polyurethane are widely used because of their availability to provide excellent adhesion to various surfaces [1, 34, 72, 109, 118]. However, along with the discrepancy between theoretical knowledge and practical application, most proposed lignin-based products failed to achieve the same mechanical properties as commercial resins or are not commercially feasible.

Nevertheless, another way has been found to open the path to the sustainable development of the new generation of lignin polymers in the form of nanoparticles and vitrimer. Vitrimer or vitrimer-like polymers are a promising alternative for thermoset applications with the advantage of recyclability [143]. In vitrimer preparation, renewable resources such as vegetal, animal, or fungal have been used widely as the building blocks with biocompatible and biodegradable properties and reduced the consumption of non-renewable resources [67, 129].

In this context, lignin provides chemical structure and unique properties that make them attractive and potentially sustainable components to vitrimer. The application of lignin as wood adhesives showed improved properties and comparable performance compared with those of commercial adhesives. This chapter aims to provide an overview of lignin-based wood adhesives by highlighting chemical modifications of lignin for recoverable adhesives.

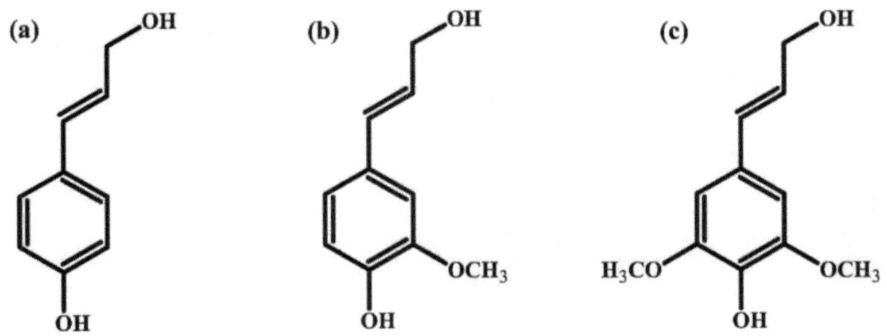

Fig. 1 Types of lignin precursors. **a** *p*-coumaryl alcohol, **b** conifer alcohol, **c** sinapyl alcohol

2 Lignin Structure and Kinds

Lignin structure

Lignin is composed of three different cinnamyl alcohol monomers: *p*-coumaryl alcohol (H unit), coniferyl alcohol (G unit), and sinapyl alcohol (S unit) (Fig. 1). The ratios of monomers differ for each species of plants. Softwood lignin contains mainly coniferyl alcohol (G unit) and hardwood lignin contains both coniferyl and sinapyl alcohol (G-S unit), meanwhile, grasses typically contain all three monomers (G-S-H unit) [28].

Functional groups are important structural characteristics in lignin because they can affect optical properties, dispersion characteristics, and chemical reactivity of lignin. Qualitative and quantitative determination of lignin functional groups has been studied for lignin structure. The functional groups of lignin consist of phenolic hydroxyl, aliphatic hydroxyl, and carbonyl which have different proportions between softwood and hardwood lignin [2]. In addition, lignin has various linkages such as β-O-4, 5–5, β-5, β-1, α-O-4, 5-O-4, and β-β as shown in Fig. 2. Research on the architectures of various lignins has been conducted using a variety of procedures and chemical approaches. These chemical units play an important role in preparing functional materials through lignin modification.

3 Technical Lignin

Lignin is the main source of phenolic compounds linked to hemicellulose and cellulose in the cell walls of vascular plants. This interlinkage has been a big challenge recently to separate lignin from biomass [32]. Technical lignin is obtained from industrial pulping processes of lignocellulosic raw materials. Therefore, the chemical structure of lignin varies depending on the origin of lignin from raw materials. Technical lignin is appreciably heterogeneous in chemical structure and molecular mass [70]. The origin and pulping condition of raw materials result in heterogeneity in aromatic units, bonding patterns, frequency, and molecular weight [23]. There are

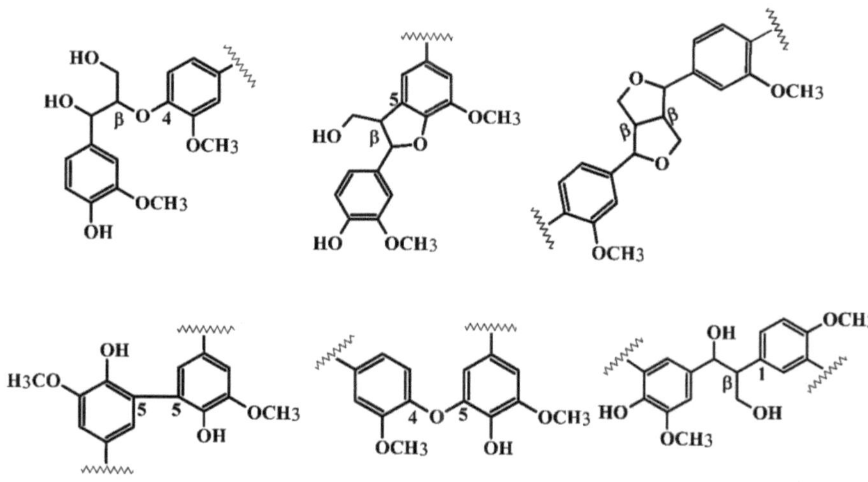

Fig. 2 Common linkages of lignin

various technical lignins from available industrial pulping processes. These technical lignins include Kraft lignin from Kraft pulping, soda lignin from soda pulping, and lignosulfonates from sulfite pulping. Different pulping processes for different raw materials result in a large variation in product purity and physico-chemical properties as presented in Table 1 [53, 117].

Table 1 Molecular weight result of various technical lignins

Technical lignin	Type of raw material	Mn (g/mol)	Mw (g/mol)	Mw/Mn	References
Kraft	Softwood	1300	3800	2.9	Wang et al. [123]
Kraft	Hardwood	1651	3041	1.9	Mun et al. [83]
Organosolv	Coconut shell	807	1598	1.98	Avelino et al. [10]
Organosolv	Softwood	2583	7747	3.00	Yu et al. [142]
Organosolv	Switchgrass	2238	6625	2.96	Yu et al. [142]
Lignosulfonate	Industrial grade	N/A	7642	4.6	Gao et al. [38]
Lignosulfonate	Industrial sample	5600	4600	1.5	Fredheim et al. [36]
Soda	Wheat straw	1067	2779	2.6	Malyar et al. [75]

3.1 *Kraft Lignin*

Alkali pulping is the most used process in industrial chemical modification which provides a high proportion of around 85% industrial lignin [23, 65, 114]. In this process, lignin eliminated from fiber cell walls is solubilized in the black liquor due to the presence of ionized phenolic hydroxyl groups that have been formed through the reaction between lignin and the pulping liquor [43, 65]. The process provided characteristic Kraft lignin by an extended 2–3 h exposure of the wood to elevated pH values and temperatures above 160 °C [23].

In the Kraft pulping process, a nucleophilic substitution reaction could happen at the β-position of the side chain which can cause degradation followed by the formation of ketone groups and carboxyl groups [52]. Characteristics of lignin from this process have a large variety of reactions that occur within the lignin structure which can create different structures of the native lignin from wood [23]. Studies of utilizing Kraft lignin have widely been done [23–25, 60, 85, 101]. A recent study showed that Kraft lignin extracted from black liquor apparently demonstrated its promising characteristics [124]. The chemical structure of Kraft lignin is shown in Fig. 3.

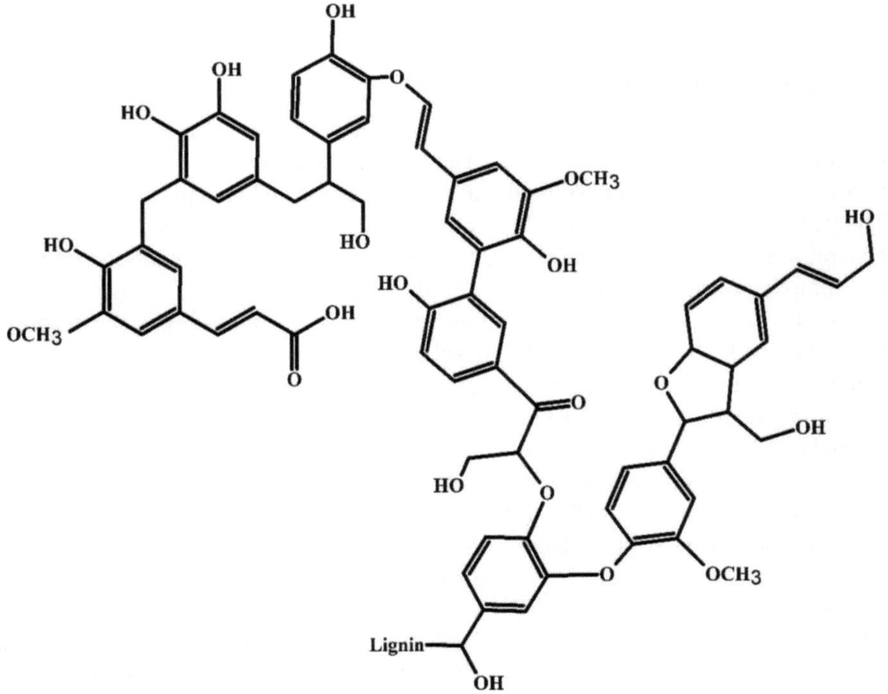

Fig. 3 Chemical structure of Kraft lignin

3.2 Soda Lignin

The utilization of soda lignin has been widely produced for making wood fibers such as straw, sugarcane bagasse, flax, etc. In addition, this process produces high-yield hardwood pulps that are intended to make packaging papers and boards [12]. Soda lignin is obtained from the soda pulping process or soda-anthraquinone. This soda lignin is getting further established since the facilities were installed first in France and India. The soda process has been known as a sulfur-free process, which is a major advantage over the Kraft operation [12, 68, 75, 120].

In the soda process, lignin contains a small number of phenolic groups while the amount of the aliphatic groups attains 2.33 mmol/g after lignocellulosic biomass extraction [49, 75]. The separation of soda lignin can be applied as Kraft lignin, yet deodorizing the liquor is not necessary. The method can be used through filtration and centrifugation of non-wood soda lignin recovery to produce high carboxylic group content [27]. On the other hand, there is a possibility of substituting wood flours with non-wood from Alpha grass soda lignin which is now a locally available by-product of the paper industry in Algeria for the separation of phenolic composites [50]. It is reported that the soda process can reduce environmental problems compared to certain extraction methods, e.g., organosolv and Kraft lignin [16, 120].

3.3 Lignosulfonate Lignin

Commercial lignosulfonate is known as an impure material and has led to improved purification methods prior to modification to obtain suitable lignosulfonate [32]. The purity of lignosulfonate might adversely affect its use in its modifications. In general, lignosulfonates contain higher sulfur, carbohydrate, and inorganic impurities compared with those of Kraft lignin [36], and its chemical structure is displayed in Fig. 4. A decline in the production and availability of lignosulfonate is leading to a reduction in the availability of lignosulfonate-based products. Therefore, chemical modification is prevented from some industrial applications owing to the high cost of the modification [32]. Lignin molecules are exposed to sulfonic acid groups throughout the sulfite pulping process because the sulfonation reaction occurs at α-position of the lignin side chain. Furthermore, condensation established α-aromatic bonds amongst lignin units [52]. It might be preferable because softwood lignosulfonate has less methoxy than those of hardwood lignosulfonate [32]. In addition, a study on the chemical structure and characteristics of sodium lignosulfonate (Na-LS) has been investigated through ion-exchange to convert Na-LS to lignosulfonic acid (H-LS) [126]. This method successfully shows greater molecular weight after ion-exchange compared with those of Na-LS and hardwood and softwood Kraft lignin [126].

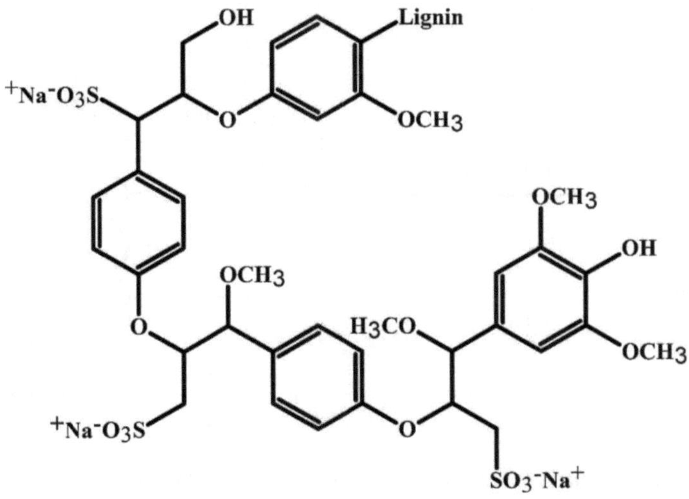

Fig. 4 Chemical structure of lignosulfonate

3.4 Organosolv Lignin

Organosolv pulping is mostly utilized in lignocellulosic biorefineries and is accepted as an environmentally beneficial substitute for Kraft pulping process. The process of organosolv lignin differs from other lignin types in conventional chemical methods. Organosolv process provides delignification which is safe and clean for producing lignocellulosic feedstock. This procedure generates lignin with high quality in the absence of sulfur, free of alkali/alkaline metals, soluble in organic solvents, etc. [107]. Organosolv lignin generated by a pulping technique involving mixed formic and acetic acids presents formulated and acetylated phenolic and aliphatic hydroxyl groups [87].

Studies have described the effect of organosolv fractionation with and without acid catalyst addition from lignocellulosic biomass with the following solvents: isobutyl ketone, ethanol, and water [53, 68]. Moreover, organosolv lignin structure has more oxidized forms due to the presence of hydroxyl and carbonyl group. It has been reported to show lower glass transition temperature (T_g) which is easier to process terminally than Kraft lignin [97]. On the other hand, ethanol organosolv process has developed and seems to be very promising among the various pre-treatment methods currently studied. This process involves sulfuric acid which acts as a catalyst to break down and modify the lignin macromolecule until it is small enough to dissolve in the aqueous alcohol solution [30]. The absence of sulfonates in the organosolv process makes it more environmentally friendly, and it possible to produce some valuable products like resins and phenol. However, there is a still limitation to produce organosolv lignin on a large scale because of the extensive corrosion of equipment and the high cost of solvent [68]. The organosolv lignin chemical structure is presented in Fig. 5.

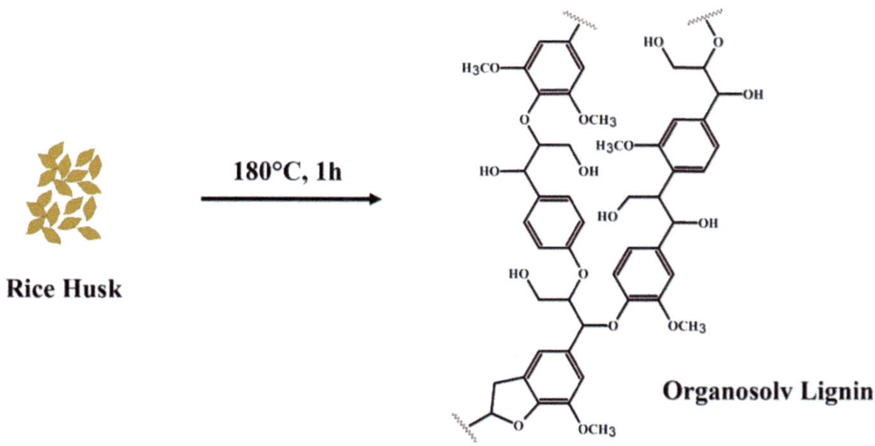

Fig. 5 Organosov lignin chemical structure extracted from rice husk (Modified from reference [107])

3.5 *Hydrolyzed Lignin*

Enzymatic hydrolysis has been studied for Hydrolyzed lignin, which is a by-product of the pre-treatment process in cellulosic ethanol plants. Based on the biorefining concept, cellulose and a part of hemicellulose can be dissolved by acid, base, or enzyme, and produce materials such as ethanol [32]. The main aim of the acid treatment method is to solubilize the hemicellulose fraction of lignocelluloses. Meanwhile, lignin remains an undissolved material and is usually used as fuel [103]. Cellulases and hemicellulases were used for the enzymatic hydrolysis of lignin to break down cellulose and hemicellulose in biomass, leaving lignin as solid insoluble leftovers that provide an economical and environmentally friendly procedure. The lignin products developed by this process showed very poor solubility in water and certain organic solvents and were produced without using sulfur. This structure is closer to native lignin than other technical lignin. Hydrolysis lignin is mainly composed of lignin up to 60% balanced with unreacted cellulose, mono, and oligosaccharides whereas enzymatic hydrolysis lignin generally contains 65–80% of lignin [74]. Other constituents like ashes, proteins, and carbohydrates are used extensively in industry to make different polymeric compounds as well as dispersants, binders, sorbents, and emulsifiers [39, 68, 96]. Hydrolysis lignin has been valorized through lignin depolymerization and has shown great promise [74].

4 Chemical Modification of Lignin

Various renewable materials such as in the form of tannins, chitosan, carbohydrates, and lignin have been used in preparing bio-adhesives for wood panels [21]. The most important variable is to improve the reactivity of lignin molecules through various chemical modifications. Chemical modification could provide some advantages of improving the reactivity and properties through dimensional stability, resistance to moisture, and biodegradation. To this end, technical lignin can be utilized in various ways to make remarkable products such as energy, liquid fuel, composite materials, and adhesives [53]. Some studies recently employed a chemical modification to solve the heterogeneity and poor solubility of lignin. The process combined physical and chemical methods that could increase the product yield of biomass [78, 95]. Some chemical modifications for technical lignin such as oxypropylation, hydroxymethylation, phenolation, glyoxalation, demethylation, allylation, etc. have been reported [14, 22, 45, 54, 57, 109]. In addition, various lignin extraction methods such as solvent fractionation, membrane filtration, sequential precipitation, microwave processing have been applied as well [15, 19, 29, 42, 95, 99, 124].

4.1 Oxypropylation

The complexity and thermal instability of structural lignin was obtained because of extreme pH and temperature treatment such as subjecting lignin to a condensed, rigid, unreactive, and insoluble material. Chemical modification such as oxypropylation has been investigated to the drawbacks of lignin utilization in polymer applications. Oxypropylation is well known as chemical modification reactions attempted on lignin which reacted with propylene oxide. Initially aimed to prepare engineering thermoplastic and polyurethane foam [18, 92, 98]. For example, the oxypropylation of agricultural waste corn stalk was modified to propylene oxide (Fig. 6), resulting in a large molecular weight of lignopolyol, the most secondary hydroxyl, and the best thermal stability. The lignopolyol encouraged the further synthesis of polyurethane materials involving hydrothermal reaction at 150 °C for 1 h [130]. Another condition required dispersion from solid lignin into neat propylene oxide at 150–200 °C [98]. Different lignins such as soda lignin, organosolv lignin, Kraft lignin, and oxidized organosolv lignin were demonstrated to oxypropylation with reaction 180 °C in the presence of KOH as a catalyst [84]. However, the optimal reaction conditions cannot be generalized to all biomasses owing to their diversity [92].

In other cases, oxypropylation was used to mask the phenolic hydroxyl group of softwood Kraft lignin under mild conditions to prevent polymerization [98]. Through this method, the phenolic hydroxyl group that is entrapped inside the molecule is free from electronic constraints where solid lignin becomes a liquid polyol [18]. Moreover, mathematical models were developed to determine responses due to the reaction efficiency of polyol yield [92]. The final product of oxypropylation, which

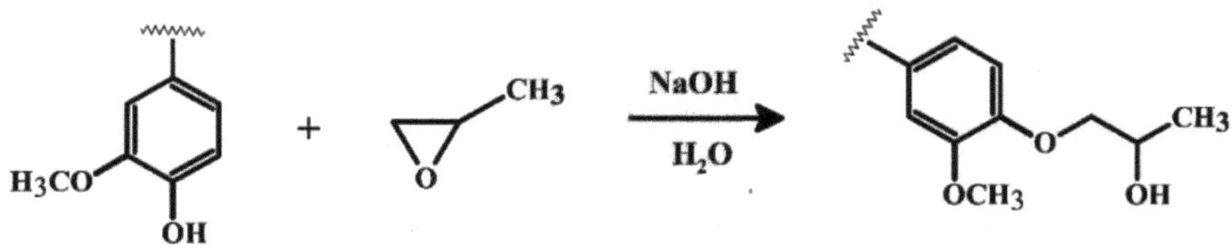

Fig. 6 Lignin structure via oxypropylation reaction

is lignin-based polyether polyols, was successfully synthesized by converting solid lignin into liquid polyfunctional polyols. It was found that there are no unreacted residues and reaction times are reasonable [18]. More final products in the form of rigid polyurethane foam (RPU) after oxypropylation showed viscous liquid, good thermal properties, and dimensional stability, which are close to those of conventional polyols used in the manufacture of polyurethane [14, 84, 100].

4.2 Hydroxymethylation

Hydroxymethylation, also called methylolation, is one of the interesting lignin chemical modifications for improving lignin reactivity due to its low cost [4, 22]. In many studies, hydroxymethylation was used to enhance the aliphatic hydroxyl of lignin and to replace phenol in lignin-phenol formaldehyde (LPF) adhesives [22]. The reaction is done by introducing the reactive position of hydroxyl groups, which notably occur in the ortho-positions of aromatic lignin as presented in Fig. 7. In addition, [47] developed hydroxymethylated lignin (HL) that is attached to *meta*-position of lignin. The reaction occurred at high temperature which led hydroxymethyl groups to react with the phenol of lignin and end up forming methylene bonds. In this situation, the cross-linked network can be formed, and the adhesive can be obtained [47].

Hydroxymethylation results in increasing the lignin content of hydrogen and oxygen by grafting formaldehyde to the phenyl propane subunits of lignin [40]. This modification was affected by several factors such as reaction conditions, including time, pH, temperature, pressure, selected catalyst, and formulation of formaldehyde and lignin. As reported, different percentages of substitution of lignin showed

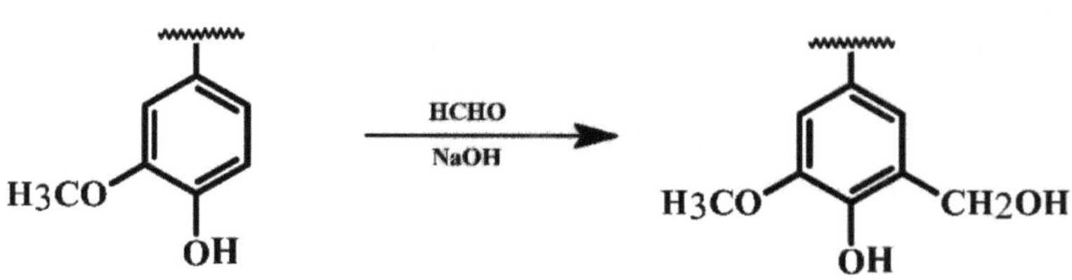

Fig. 7 Hydroxymethylation of the phenolic ring of lignin

different qualities and properties of adhesives [22]. However, the reactivity or cross-linking depends on the type of wood from pulping conditions whether it is softwood, hardwood, or grass.

In recent studies, HL from softwood lignin can replace phenol by 35% for wood panel adhesives production [4]. Moreover, the organosolv lignin can substitute up to 50% of the phenol with desired properties [13]. Nevertheless, the formaldehyde emission from the final product bonded with PF adhesives was something concerned about a risk to health [22]. In other studies, HL produced sulfonate with sulfuric acid or sodium sulfite under different conditions. The result showed that the modification increased HL molecular weight compared with those of Kraft lignin. Both HL and Kraft lignin had sulfur that was impurities in pure Kraft lignin as pure Kraft lignin was washed with sulfuric acid prior to use [40]. HL is also expected to form a denser crosslinking network with the hydroxyl group of isocyanates for improving lignin-based polyurethane. It has been proven that lignin reactivity was effectively enhanced by hydroxymethylation [40].

4.3 Phenolation

The toxicity, formaldehyde emission, and high cost of PF have encouraged studies to produce green-based panels. Phenolic modification of lignin is the common modification method of lignin to prepare modified lignin-phenolic resin by substituting formaldehyde and phenol. This modification generates modified PF resins that could reduce economic costs and improve the economic value of utilizing lignin [20, 41, 48, 52]. Two commercial lignins were investigated for a comparison of phenolation. As shown in Fig. 8, pine Kraft lignin and sweetgum biorefinery lignin were modified to improve lignin reactivity by phenolation [57].

A study modified lignin to make phenolic impregnation resin. After involving with one of two different lignins, the phenolated lignin had lower molecular weight compounds below 500 kDa [115]. Synthesizing lignin with phenol means to reduce phenolic content and curing temperature [52]. However, the limitation of substituting technical lignin in PF resin could be overcome by enhancing the reactivity of lignin by increasing the content of methylol, phenolic hydroxyl, or alcoholic hydroxyl. This

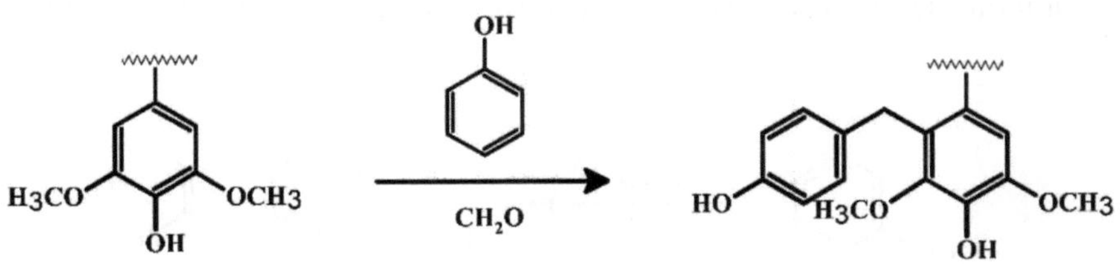

Fig. 8 Phenolation mechanism of lignin

method may enhance the polymerization degree of lignin-based-formaldehyde and can improve the resultant strength in producing composite wood panels [41, 48].

In addition, the phenolation treatment involving alkaline conditions can effectively increase the lignins, especially in the active sites which can improve the rate of synthesis process of LPF resin. The comprehensive treatment for technical lignins provides promising results for further utilization in producing LPF resin adhesives [135]. Moreover, another modification can be added to improve phenolation reaction by employing an acidic catalyst in the reaction process. The process aims to form a network bonding of phenol at the *ortho* or *para* position of the lignin phenolic hydroxyl group, leading to making rigid material at the certain temperature [41].

4.4 Glyoxalation

Environmental concerns have been encountered in building eco-friendly adhesives for wood bonding. The toxicity of formaldehyde encouraged several researchers to substitute the use of formaldehyde in wood adhesive, including glyoxalation. Like other modifications, glyoxalation was done to enhance the reactivity of lignin, using glyoxal, a non-volatile aldehyde. Unlike formaldehyde, glyoxal has less reactivity yet it has been used in lignin for application to wood panels such as particleboard [87]. The glyoxalation method could be a proper way of improving the reactivity of native lignin to show efficient function in the wood adhesive [21]. In other words, nontoxic wood adhesives such as low formaldehyde emission can be obtained by applying glyoxalation through substituting formaldehyde with glyoxal as presented in Fig. 9. It is proven that low molar ratio of formaldehyde could improve high formaldehyde emission of environmental concerns from wood panels [139].

In consequence, glyoxalation methods have been reported in a wide range of studies. A novel study has developed self-crosslinking by glyoxalated lignin to form a three-dimensional network sample for replacing formaldehyde-based resin [45]. Another study has modified urea cross-linked glyoxylated lignin (ULG) with dialdehyde [21]. This method involves two materials such as glyoxalated lignin and hydrogen peroxide to be crosslinked with urea. In addition, glyoxalated lignin has been applied to produce green adhesives for wood panel applications. The adhesion performance of substituted formaldehyde with glyoxal was comparable with lignin-formaldehyde and conventional adhesive [106]. However, a lack of reactivity

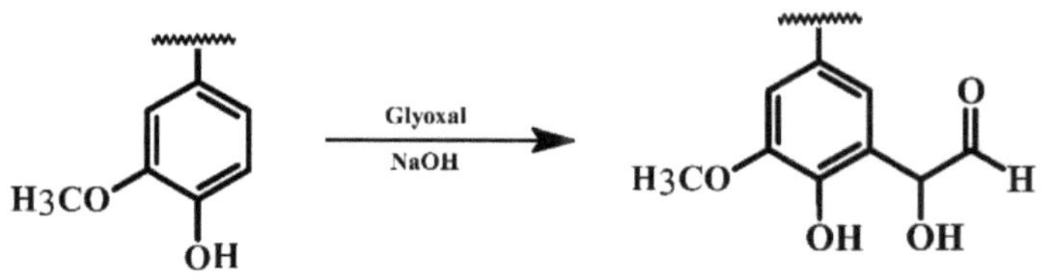

Fig. 9 Glyoxalation mechanism of lignin

of glyoxal and lignin has led to lower strength and thermal properties of ULG resin. Thus, an additive is required to solve this problem.

Furthermore, the use of glyoxalated lignin has improved in the creation of carbon material. With polyacrylonitrile/Kraft lignin (PAN/KL), the morphology of the glyoxalated lignin has been conducted [139]. To produce potential carbon material, Kraft lignin is necessary for the electrospinning process, during which PAN/KL is recovered as nanofibers. The glyoxalated lignin has improved the dimensional stability of the anti-swelling efficiency of Jelutong wood. As a bulking agent, the resin of glyoxalated low molecular weight lignin phenolic resin can be used in several applications in the form of parquet flooring, paneling, and furniture components [7].

Another study has proved that lignin isolated from coconut husk could be green material through lignin glyoxalation in the production of wood adhesive [11]. It showed that 30% lignin substitution of phenol-glyoxal generated remarkable tensile strength and MOE. Moreover, the addition of layered double hydroxides as reinforcement shows better physicochemical properties of glyoxalated lignin-phenol. The gyloxalation process revealed that the cross-linking bridge was done through the formation of ether linkages [45].

4.5 Demethylation

The functional groups of lignin that are similar to phenol are promising substitutes for phenol in making lignin-based adhesives. Another advanced modification has been found to increase the reactivity of lignin due to its large molecular weight as a complex structure. Demethylation of lignin by utilizing of aryl methyl ethers has been demonstrated using iodocyclohexane with removing methyl group under mild conditions [151]. This result shows a new approach to developing lignin-based adhesives. In addition, lignin demethylation has been found to work by involving enzyme reactivity. This reaction involved with lignin-degrading fungi such as brown-rot, white-rot, and soft-rot fungi has not yet been developed. This kind of unusual phenomenon of lignin demethylation was done by removing the O-methyl/methoxyl groups from lignin with enzyme activity. In this observation, Kraft lignin played a key role in the biological demethylation process where carbohydrate substrates were added to nutrient media [119].

Challenging lignin demethylation to increase reactivity is an important application due to increasing enhanced crosslinking bonding through hydroxyl group as shown in Fig. 10. The demethylated lignin obtained from wheat straw alkali lignin was demethylated and was used to substitute 60 wt% of phenol to prepare LPF. The triumph of demethylation is evidenced by shorter gel time, lower free formaldehyde, free phenol contents, and higher bonding strength compared with without demethylation process [109]. Moreover, another enhanced reactivity of Kraft lignin can be done by using lithium bromide trihydrate. The depolymerization and demethylation of Kraft lignin is done by cleaving aryl and alkyl ether. The polymerization via β-O-4 cleavage occurred under mild conditions while sufficient demethylation of the lignin

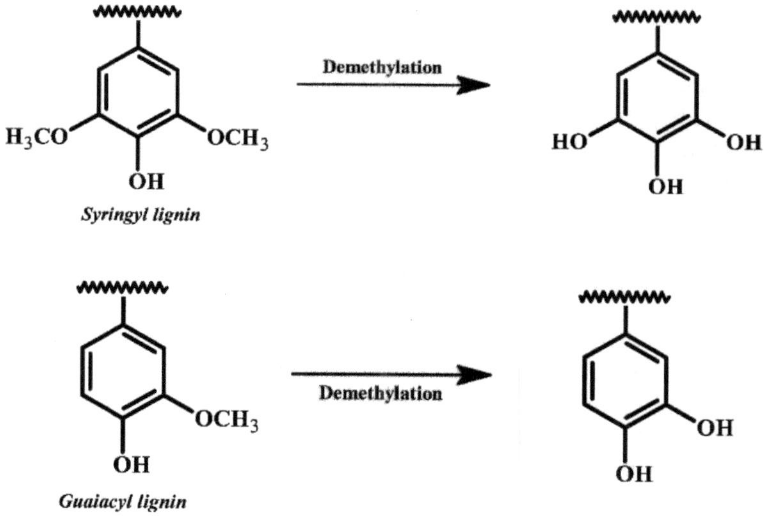

Fig. 10 Demethylation mechanism of lignin

needed harsher conditions. However, a new aromatic hydroxyl group (Ar-OH) has occurred after the polymerization and demethylation [136]. Demethylation can also occur from laccase treatment by releasing methanol because of the methoxy group splitting from the aromatic rings and decrease of methoxy content. This method is considered to form catechol structures [121].

4.6 Allylation

Allylation is a new advanced approach that utilizes hydroxyl groups and provides a versatile product [88]. Lignin allylation converts lignin into a thermoset polymer as a promising route for lignin utilization. This method is reported to be an efficient, non-toxic, and sustainable modification technique with various thermoset properties of different lignin fractions (Table 2). For instance, the allylation of coniferyl alcohol under ethanol shows a remarkable selectivity toward phenols in the presence of aliphatic hydroxyls as shown in Fig. 11 [56, 95].

Table 2 Lignin allylation of various sources of lignin

Source	Allylated material	D	References
Kraft lignin	Allyl chloride	2.5	Jawerth et al. [55]
Organosolv lignin	Tetrabutylammonium bromide	3.1	Over and Meier [88]
Softwood lignin (TMP)	Allyl bromide	2.3	Zoia et al. [149]
Herbaceous lignin (P1000®)	Allyl bromide	1.8	Zoia et al. [149]

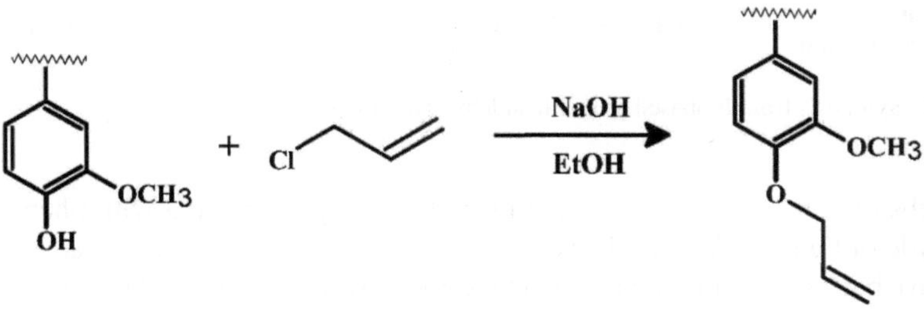

Fig. 11 Allylation of coniferyl alcohol. Reproduced from Ref. [55] with permission from the Royal Society Chemistry

In addition, an allylated lignin was obtained through Claisen rearrangement which is a simple reaction to introduce carbon–carbon bonds and release phenolic groups [149]. In this reaction, purely guaicylic lignin (TMP) and herbaceous lignin (P1000®) were taken as reference materials. The investigation was focused on the lignin structure affection and the reactivity composition. Another study reported that aromatic and aliphatic hydroxyl groups were allylated up to 90% using diallyl carbonate as a reagent at relatively low temperatures (90 and 120 °C) of beech wood organosolv lignin [88].

In addition, the allyl group introduced during the allylation is considered as a possible derivation route to generate versatile chemicals. Meanwhile, the dialkyl carbonate utilization has shown promising results in the form of phenolic building blocks of lignin from allylation modification which is sustainable and environmentally friendly. Allyl ether was used as a route to selectively modify lignin phenols [54]. Moreover, it is possible to modify the polymer via highly efficient thiol-ene reactions with a variety of thiols [144]. Thiol-ene is one way to utilize the allylated lignin by covalently coupling two molecular entities together. This reaction is considered as a new path with versatile tools for modification and cross-linking of ene-containing moieties [55].

5 Lignin-Based Thermosets

5.1 Lignin-Based PF Resins

PF resins are polymeric substances that can be produced by reacting phenol with formaldehyde in either an acidic or basic environment [140]. This resin is one of the most used adhesives for wood with high performance such as strength and moisture resistance. However, there are limitations to using wood-based panels because of phenol toxicity, formaldehyde emission, and the cost of PF [140]. As explained before, environmentally friendly phenolic resins for wood-based panels can be realized by phenolic modification. This modification is most applicable and potentially

Aromatic model compound + **Phenol** ⟶ **Resol prepolymer** ⟶ **Cured phenolic resin**

Fig. 12 Synthesis from biobased aromatic aldehyde compounds

can substitute the formaldehyde and phenol, leading to reacting with phenols and aldehyde of lignins (Fig. 12). Hence, various types of lignin extraction are required and have been widely investigated to make good raw materials for phenol production [62]. It is no wonder that this modification has generated resins with low economic cost and improved the economic value of lignin [2, 48, 52].

The advantages of LPF resin synthesis are to reduce phenolic content in free formaldehyde and free phenol, and to reduce the curing temperature [52]. Nevertheless, technical lignin has hindered the optimum condition of substitution phenol in LPF due to lower reactivity compared to phenol [121]. To this end, some modifications for the lignin structure have been done to increase the content of methylol, phenolic hydroxyl, or alcoholic hydroxyl [41, 48, 62]. This method may enhance the degree of polymerization of LPF resins which can improve its resultant bonding strength in producing composite wood panels as presented in Table 3.

Lignin molecules have the privilege to provide phenol structures that have strong reactivity and can undergo the condensation reaction of phenol and formaldehyde under alkaline conditions. The phenolic mechanism of lignin can be seen in Fig. 13. Under these circumstances, the electron-induced action of phenolic hydroxyl groups can only break the α-ether bond or double bond of the side chain in phenolic lignin, forming the methylene quinone structure. On the other hand, phenol can react with the abundance of aldehydes in lignin structural units under an acidic environment. This state is brought about by the ether bonds at the α-position of the lignin-phenylpropane

Table 3 Properties of lignin-based PF resins

Sample	Solid content (wt%)	Viscosity (mPa. S)	Free formaldehyde content (%)	References
Lab PF resin	30.0	n/a	0.30	Arefmanesh et al. [8]
Kraft lignin	20.4	n/a	1.30	Arefmanesh et al. [8]
Kraft lignin	20	110	2.40	Kalami et al. [59]
Kraft lignin	67.9	n/a	n/a	Thébault et al. [115]
Kraft lignin	36.0	61	0.20	Ma et al. [73]
Soda lignin	n/a	415	n/a	Younesi-Kordkheili [138]
Alkali lignin	49.9	243	0.91	Lee et al. [64]
Hydrolysis	42.7	33.2	0.91	Paysepar et al. [90]
Lignosulfonate	65.9	n/a	n/a	Thébault et al. [115]
Lignosulfonate	50.8	264	0.57	Lee et al. [64]
Organosolv	43.9	133	0.24	Pang et al. [89]

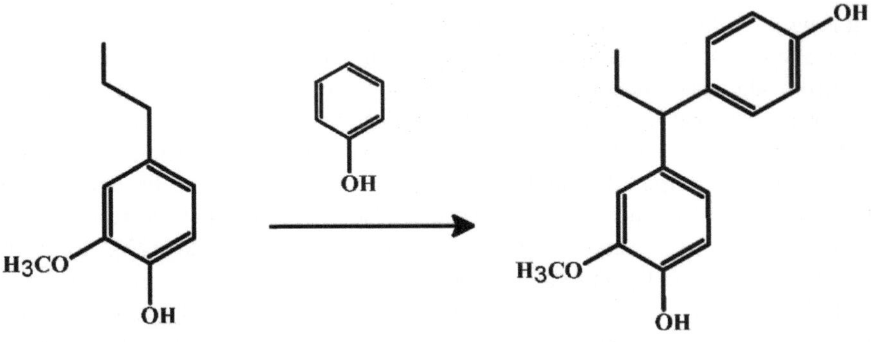

Fig. 13 Phenolic mechanism of lignin

structural unit being easily broken and forming oxonium by a combination with hydrogen units. The phenolic alteration may result in a decrease in lignin's molecular weight and polydispersity [48, 52].

The utilization of lignin in biomass-based phenolic resin has been widely investigated. An analysis of the modified resins was created by condensing lignin, phenol, formaldehyde, and whisker silicon to increase the heat resistance and thermal stability of phenolic foam. Lignin and whisker silicon addition have an impact on the foam's flame retardancy, heat release, preservation, and thermal stability [145]. Another investigation prepared with phenolic oils derived from lignin can effectively improve the cost and environmental friendliness of polymers. This analysis focused on the multi-spectroscopic synthesis of lignin-based bio-phenolic resin. The molecular chain structure of lignin-based phenolic resin can be disordered by adding 50 wt% of phenolic oil. In the meantime, formaldehyde rather than phenol was the preferred polymerization partner for the active phenolic oil. By using lignin and phenol, phenolic molding plastic materials were produced that were more stiff and thermally stable, providing a bio-based phenol molding polymer materials that were highly effective and environmentally friendly [20].

A study investigated modified LPF panels bonded with maleic anhydride, mentioned as maleated lignin. The addition of maleated lignin in PF resin with a high proportion reaching 50wt% could decrease costs and toxicity . However, the hydroxyl group content decreased due to the treatment of maleic anhydride. Nevertheless, it was found that the panels bonding with maleated lignin showed early curing time, lower formaldehyde emission, and lower water adsorption. The increase of maleated lignin from 30 to 50wt% improves the panel's mechanical strength, dimensional stability, and formaldehyde emission as well [140].

Utilizing Kraft lignin as raw material is such a challenge to build proper raw material fraction. Lignin has a closer backbone structure to phenol thus more suitable for replacing phenol even 100% in phenolic adhesive formulations [59]. A lignin modification in the form of fractionation with organic solvents has been studied. Acetone fractionation is one of the potential methods to create high-potential raw material lignin [124]. They showed that this technique allowed a successful synthesis of LPF resins with soluble and insoluble Kraft lignin fractions with a range of chemical and physical characteristics. The soluble Kraft lignin obtained by acetone fractionation resulted in abundant phenolic hydroxyl units which is crucial for the LPF synthesis. However, hardwood Kraft lignin with mostly high syringyl groups and two occupied methoxy groups in the *ortho* position is not reactive toward formaldehyde. Softwood lignin which is composed of high guaiacyl content gives a chance to replace with 100% phenol in LPF formulation [59]. It is also confirmed that the reactive site (C5 *ortho* position) is available for the reaction with formaldehyde [127]. However, acetone-soluble Kraft lignin from softwood and hardwood has the potential to replace phenolic components. It has low polydispersity, more condensed lignin structure, and no carbohydrate impurities [128]. The synthesis of LPF resins with acetone-soluble fractionated softwood lignin has been widely studied to replace phenol in LPF resins. In fact, it resulted in higher adhesion performance compared with those of unfractionated Kraft lignin [8, 59].

An important thing that should be noticed to make LPF is the composition of phenol/formaldehyde molar ratio, catalyst concentration, and amount of lignin substitution for the phenol. It is reported that lignin-modified phenolic lignin is more stable than the resol (pure phenol resin). However, the adhesive strength decreases as the lignin content increases [35, 105]. A distinct method of creating formaldehyde-free resol phenolic resins has been developed. It involved functionalizing phenolic compounds with aliphatic aldehydes, which involved deprotecting acetal groups into aliphatic aldehyde groups after they had been functionalized on an aromatic hydroxyl function into two phases. Brominated reagents were needed to complete the procedure although they could be hazardous when applied [35].

5.2 *Lignin-Based Epoxy Resins*

Many efforts have been directed to use more reliable lignin feedstocks in the synthesis of lignin-based thermosets to overcome the limitation of lignin molecular structure. In this sense, lignin has been used to synthesize several resins in the form epoxy resins. The epoxy resin is an important and large market with specific reference to phenol-epoxy resins [76]. The type of lignin and parameters such as the chemical composition of raw material, skeletal structure, and functional group distribution, cross-linking agent could directly affect the material properties of modified epoxy resin [46, 52].

In lignin-epoxy resins, lignin can improve the properties of epoxy resins such as thermal stability and solvent resistance, leading to an optimum strength and flexibility of the resin. Lignin-modified with methanol is encouraged to replace bisphenol A as a raw material for epoxy resin preparation due to susceptible biodegradation materials [86]. The process of based-epoxy resins started with increasing the phenolic group content in lignin. It is commonly phenol and bisphenol A which are supposed to react with ECH and lead to epoxidized lignin as shown in Fig. 14. During modification

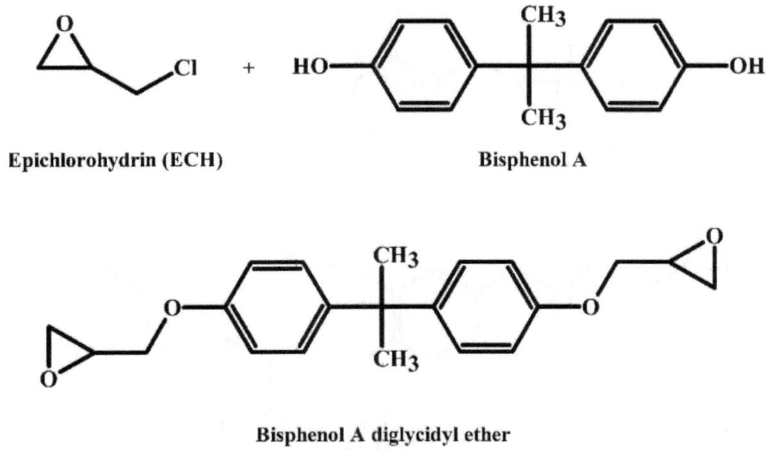

Fig. 14 Schematic reaction of epoxy resins

of lignin-based epoxy resin, cross-linking reactions could involve polyamine curing agents. This method provided better waterproof adhesion strength [9].

In addition, there is a correlation between molecular structure and properties of lignin-based epoxy resin. It showed better thermochemical properties of softwood spruce Kraft lignin compared with eucalyptus-based resin [46, 118]. Adhesion in lignin-based epoxy is also important. The strength depends on the chemical composition, synthesis procedure, and substitution degree of the adhesive with bio-epoxy [118]. The schematic of adhesion strength measurement is shown in Fig. 15.

A study investigated the formation of Kraft lignin-based epoxy resins by the epoxidation reaction with lignin modification of methylolation reaction. The methylolated lignins were obtained by reaction of original Kraft lignin with formaldehyde and

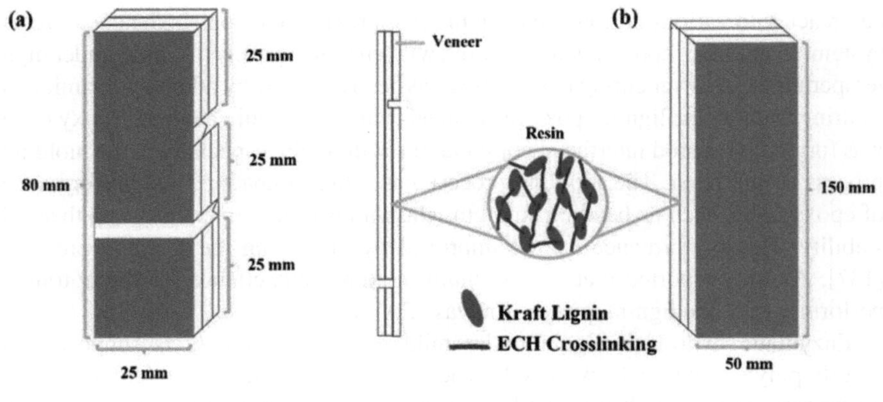

Fig. 15 Schematic specimens **a** TSS measurement and **b** FE measurement of lignin-based resin crosslinked with epichlorohydrin [125] (Permission from Taylor and Francis)

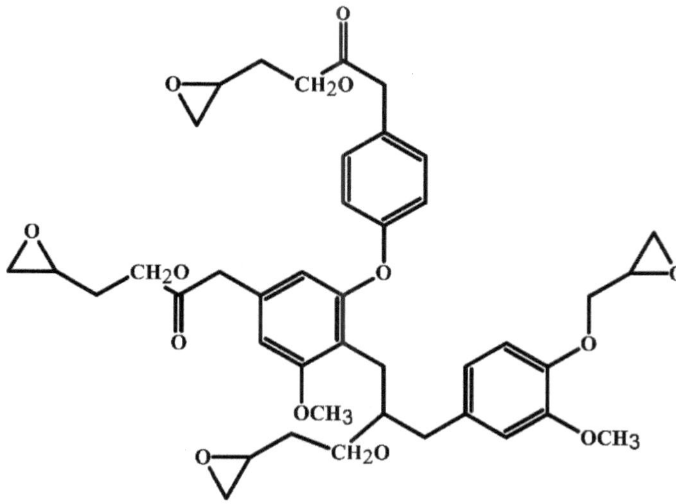

Fig. 16 Lignin-based epoxy structure mechanism

glyoxal under alkaline conditions [76]. An identifying lignin-based epoxy composition of sufficiently high T_g has also been investigated. The development of epoxy adhesives was reported to build urea-lignin-glyoxal (ULG) resins. It has improved solid content, density, viscosity, and gel time. The plywood adhesion with this resin revealed improved dry and wet shear strength as the amount of epoxy increased, whereas their water absorption decreased. However, the curing temperature is lower than that without epoxy addition [139]. The mechanism of lignin-based epoxy resin is served in Fig. 16.

The excesses of epoxy resin consist of good adhesion, small curing shrinkage, good heat resistance, excellent solvent resistance. Epoxy resins have a wide range of applications such as adhesives, paints, and flux fields [9, 76]. Fortunately, lignin can react with various curing agents in the crosslinking reactions in the epoxy resin system. In addition, epoxy resin was cured with enzyme hydrolysis lignin under high temperatures. A lower curing temperature has been achieved by adding polyamine as a curing agent to the lignin-epoxy resin blends with 60% lignin content. Epoxy resin was formed by a good interfacial combination with lignin, especially in the molding process of the resin. The molding process has a high impact on the performance of epoxy resin, such as bending strength, glass-transition temperature, and thermal stability. This performance can be improved by increasing the molding pressure [137]. A study reported that the optimum molding temperature for the optimum performance of the lignin-epoxy resin was 130 °C.

Enzymatic hydrolysis lignin under mild conditions process is suggested to modify polymer materials by directly reacting with formaldehyde, epichlorohydrin, isocyanate, and so on. For example, enzymatic hydrolysis lignin from corn straw lignin with lignin content up to 60% was blended with thermosetting epoxy resin. It shows the reduced viscosity as the molding temperature increases [137]. Unlike enzymatic hydrolysis lignin, bisphenol-A was mixed with sulfate lignin with an additional high activity of aliphatic polyamine as a curing agent. The results revealed that the epoxy resin network had been achieved by synthesizing lignin with an unreacted amine group as a curing agent. The optimum performance such as compatibility and bond strength was achieved with 20% lignin in mass fraction. However, the shear strength decreased gradually when the lignin mass fraction exceeded 35% [33, 52]. The aldehyde from epoxy showed a promising mechanism to synthesize with technical lignin and make lignin-based epoxy resin.

5.3 Lignin-Based Polyurethanes

Lignin-based polyurethane is a common approach for creating value-added utilization. Polyurethane is created by aromatic polyols crosslinked with isocyanates as shown in Fig. 17. A copolymerization of lignin-containing synthesis forms polyurethane, where lignin serves as the hard segment in the polyurethane structure. Lignin hydroxyl groups in this case, readily react with isocyanates to form polyurethanes. In the present study, most of the investigated lignin-derived polyurethanes are thermosets highly cross-linked or network polymer and are barely reprocessable or recyclable [102]. Wang et al. [122] reported a synthesis of polyurethane adhesives by employing the reaction between lignin and secondary polyols with three different fractions which were high, medium, and low molecular weight. They used softwood Kraft lignin through acetone-methanol-hexane cosolvent precipitation, and related the molecular weight of lignin to the mechanical properties of polyurethane adhesives. An increase in the molecular weight of lignin improved the stiffness or resistance which was more flexible and tougher of lignin-based polyurethanes than those of the control. However, the addition of lignin up to 30–35% can cause the networks of lignin-polyurethane too brittle to measure regardless of the NCO/OH ratio applied [122].

Fig. 17 Reaction scheme of polyurethane resin

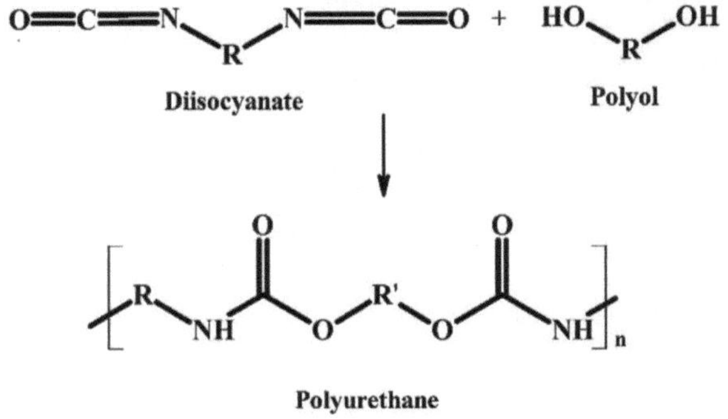

Another investigation provided an improved strategy to form robust materials from lignin by synthesizing polyurethane thermoplastics [102]. However, high polydispersity of lignin molecules played a role in hindering lignin thermoplastic copolymers from synthesizing high-performance lignin. A reaction between lignin content (65–80 wt%) and formaldehyde in mild conditions led to a successful formation of high molecular weight lignin with high yield and controlled T_g with diisocyanates as a rubber as shown in Fig. 18. However, lignin-containing recyclable polyurethane thermoplastic has not yet been developed [102]. An unprecedented investigation reported that Kraft lignin was used as a reagent in the synthesis of lignin-based polyurethane film. The synthesis of polyurethane from Kraft lignin with high purity and castor oil has been successfully demonstrated under mild conditions at room temperature. The result shows that 30% Kraft lignin improved tensile strength and moduli, and 50% Kraft lignin was good at forming polyurethane films compared with the control film form castor oil as a polyol. The substances provided potential in various applications such as coating and packaging [17].

In recent studies, the development of isocyanate polyurethane resin has been encouraged by utilizing non-isocyanate polyurethane (NIPU) which is less toxic and non-dangerous to make wood adhesives. NIPU resin has been prepared and characterized from organosolv lignin as raw material. The urethane linkage indicates that all –OH groups of both aromatic and aliphatic lignin reacted with DMC and amine. This resin has proven the satisfactory adhesion performance [104]. Another study

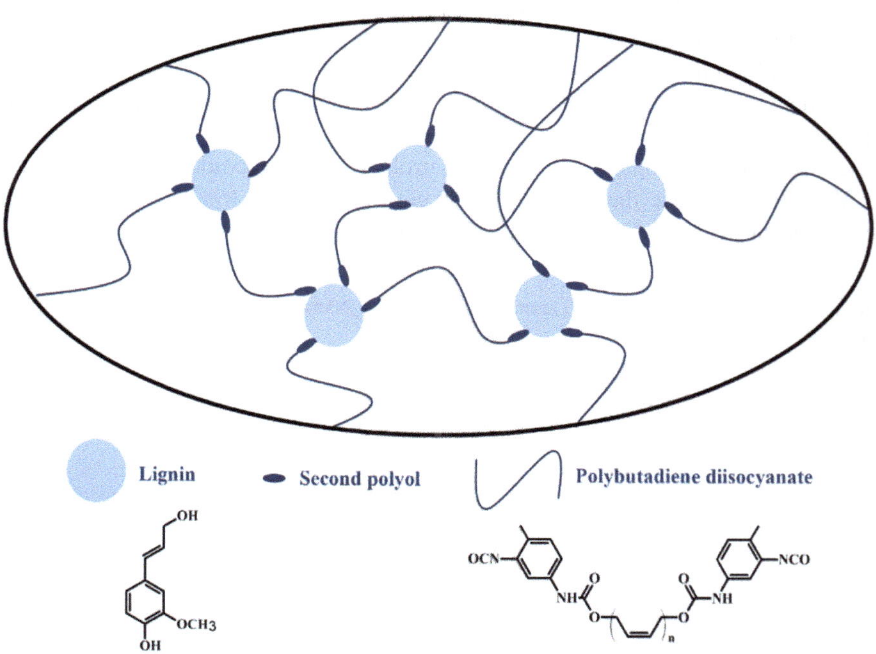

Fig. 18 Synthesis of lignin polyurethane with diisocyanate

observed the properties of NIPU thermosets with the effect of structural features in lignin in the form of methoxy content and bridging carbon substitutions [80]. Moreover, the synthesis of NIPU and polyethramine has been investigated with outstanding adhesion strength of up to 24.5 MPa [146]. These novel approaches show great potential to prepare lignin-based NIPUs materials.

6 Lignin-Based Nanoparticles

The utilization of industrial lignin has shown benefits in both economic and environmental aspects. With the excellent properties and extensive applications of nanomaterials in various fields, lignin has been shown as a possible material to prepare aqueous lignin nanoparticles. In recent years, lignin-based nanoparticles have been introduced to different kinds of lignin modification. Self-assembly, solvent exchange, acid precipitation, polymerization, ultrasonication, crosslinking, CO_2 antisolvent, etc. are common modifications of preparing lignin-based nanoparticles [111]. The modifications of lignin nanoparticles prepared from technical lignin such as alkali, Kraft, organosolv, lignosulfonate, and enzymatic hydrolysis lignin. For example, lignin lignosulfonate investigated the possibility of forming into reverse micelles because it has strong hydrophilicity which is of great significance for both the utilization of renewable resources and environmental protection [111, 112]. In addition, acid precipitation of lignin has been demonstrated by adding hydrochloric acid into ethylene glycol solution of lignin followed by cross-linking and water dialysis, which could keep stable at a wide range of pH. This extraction can introduce hydrophilic groups such as aliphatic and phenolic hydroxyl, carboxyl, and sulfonic groups. Therefore, technical lignin in the form of alkali lignin and lignosulfonate are natural polymeric surfactants [147].

Pickering emulsion stabilized with lignin nanoparticles has also been investigated. In this context, colloidal lignin nanoparticles (CLPs) would act as amphiphilic of incorporated hydrophobic polymer and hydrophilic cellulose nanofibrils (CNF). The CLPs would be able to form non-covalent bonds such as hydrogen bonding with both CNF and hydrophobic polycaprolactone (PCL) together. This strategy resulted in composites with higher tensile strength and toughness compared to pure CNF nanopapers, whether in dry or wet conditions [61]. Another study produces CLPs with phototoxic and photoallergic responses that are associated with traditional UV filters with three different technical lignins. The result showed high UV absorbance capacity obtained by Kraft lignin CLPs from dimethylisosorbide (DMI) nanoprecipitation, leading to direct application in sunscreen formulation (SPF 15, 30, and 50+) [91]. In addition, another study has involved heat treatment to produce CLPs from nanocellulose fibrils (CNF) of laccase-treated lignin. The modification of such adhere chamois and CNF surface has shown potential [77].

A study developed water-dispersed lignin nanoparticles (LNPs) that could have many applications such as bio-based adhesives, stabilizers, UV-blocking additives for thermoplastics, drug delivery, and Pickering emulsions [69, 112]. Two types of LNPs have been prepared through self-assembling and are incorporated with poly(vinyl alcohol) (PVA) to produce nanocomposite films, from deep eutectic solvent (DES) and ethanol-organosolv lignin. It was found that shells of LNPs have abundant phenolic hydroxyl groups, which induce good interfacial adhesion with the PVA matrix. To this end, it has improved the mechanical and thermal performances of the nanocomposite films [116]. In advanced research, spherical lignin nanoparticles (LNPs) have been prepared to produce lignin-epoxy hybrid nanoparticles (hy-LNPs). The hy-LNPs with $\geq$30 wt% of Bisphenol A diglycidyl ether (BADGE) as crosslinker, generate high potential bio-based waterbone adhesive formulation due to inter- and intraparticle crosslinking at >150 °C. Besides, the dry/wet adhesive strength showed a competitive number (5.4/3.5 MPa). In addition, both BADGE and lignin have substituted benzene rings that could be coprecipitated with softwood Kraft lignin to form hy-LNPs [150]. However, LNPs cured with epoxidated Kraft lignin obtain high shear strength (above 13 MPa) and very good wet strength. The epoxidation reaction happens using epichlorohydrin from glycerol and is then cured with LNPs to produce composite adhesive for wood panels [51].

7 Lignin-Based Vitrimer

Bio-based polymers have been considered as answers for sustainable development to replace fossil-based materials. These polymers refer to materials derived from or composed of biological issues from biomass including plants, animals, mushrooms, and different micro-organisms [71]. Based on this issue, there has been growing obvious development of biobased vitrimer owing to the lucrative properties and relatively high abundance of low-cost raw materials [26]. Leibler and co-workers first demonstrated the synthesis of the polymer with a concept called vitrimer by using simple carboxylic acid-based transesterification reactions, promoted by a catalyst. In addition, an approach was performed to analyze the evolution of the epoxy vitrimer network structure through statistical analysis during transesterification reactions. Transesterification acts as a disproportionation reaction that converts two linear fragments into a terminal fragment and a branching fragment. This reaction offered the ability to improve crosslinking concentrations and pendant chains while leaving a sol reaction. A transesterification catalyst is suitable for epoxy/acid or epoxy/anhydride polyester-based networks in the associative of covalent adaptable networks (CANs). These CANs represent a trend in polymer network fabrication toward the rational design of structural materials possessing dynamic characteristics for specialty applications [62, 63]. In addition, it can feature dynamic covalent bonding at crosslinking sites, seeking to combine the processability advantage that is inherent to thermoplastics with the mechanical robustness of conventional thermosets [5]. CANs consist of dissociative CAN and associative CAN which are divided based on the exchange reactions (Fig. 19). The vitrimer belongs to the group of associative CAN [26].

In addition, transesterification reaction with thermal conditions shows permanent polyester/polyol networks which have gradual viscosity upon heating. This process

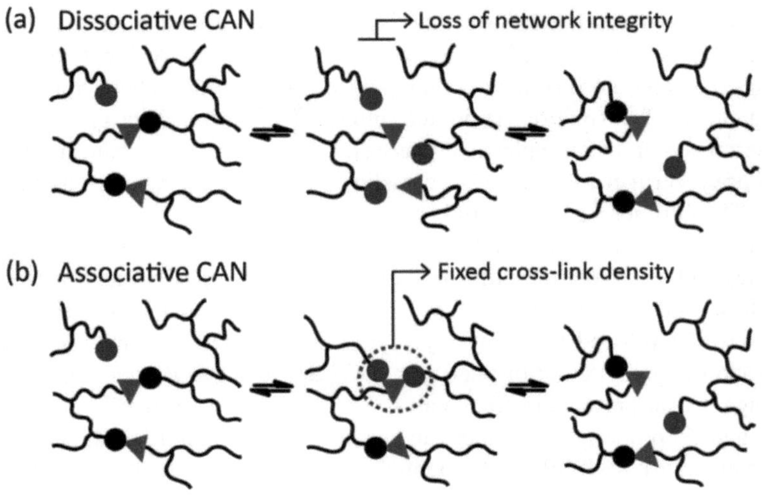

Fig. 19 Schematic of **a** Dissociative CAN and **b** associative CAN or vitrimer. Reproduced from References [26] with permission from the Royal Society Chemistry

provides vitreous silica that has never been reported in organic materials [6, 26, 81]. With the presence of a zinc catalyst on transesterification, this vitrimer system can achieve rapid stress relaxation by undergoing a transformation of network topology. Smallenburg et al. [108] proposed a concept of vitrimer to study the phase behavior and dynamics of vitrimers. The simple vitrimer model shows direct insight into the physics involved in their quality behavior and serves the system to arrange its network topology at high bond-switching rates which correspond to high glass transition temperature (T_g) and catalyst concentrations. This corresponding mechanism, makes the possibility of mimics to repair or recycle vitrimer at high temperatures and directly captures the strong glass-forming ability shown in vitrimer [108]. In addition, vitrimer have a gradual change which makes them recyclable and malleable according to their temperature following the Arrhenius law, compared to thermoplastic polymers with a sudden drop in viscosity near the glass T_g [148]. The recyclable vitrimer has comparable mechanical properties and solvent resistance, and provides a new modification method known as vitrimerization. Vitrimerization is addressed by converting the permanent cross-linked thermosets into vitrimer polymer without polymerization.

The biobased-vitrimer has been developed in recent years to overcome the recycling and reuse problem for traditional thermosetting materials [132]. Chemical modification is needed to generate synthons able to react, form dynamic covalent networks, and use a catalyst to trigger the reversibility of dynamic covalent bonds [26]. The existence of dynamic covalent bonds in crosslinked network structures of vitrimer, can solve the insoluble and infusible problems of traditional epoxy resins after curing [81].

Vitrimer can flow and behave like viscoelastic liquids at high temperatures, therefore they are reprocessed and recycled as thermoplastic. At low temperatures, exchange reactions are very slow, resulting in classical thermoset behavior [129]. An investigation has been reported to produce biobased lignin-based epoxy vitrimer with both thermosets and thermoplastic properties. Commonly the mechanism of lignin-based epoxy vitrimer involves other crosslinkers to build the network as shown in Fig. 20. The vitrimer was successfully prepared from fractionated lignin and sebacic acid without additional chemical modification. This resulted in excellent mechanical properties and self-repairing efficiency of 96.57% and reprocessing features with the highest total tensile strength of 39.82 MPa [131]. Another lignin-based epoxy vitrimer was developed involving glycidyl etherified enzymatic hydrolysis lignin (GEL) and diglycidyl ether bisphenol A (DGEBA) using dodecanedioic acid as a curing agent in the presence of zinc catalyst. By changing the ratio of GEL to DGEBA, the mechanical property of vitrimer can be adjusted. In addition, the highest tensile strength can reach 46.8 MPa with the highest replacement rate of LEVs system at 100% and the lignin content is 72% [132]. The resultant vitrimers are believed to have great potential for producing packaging materials with light-sensitive commodities [113].

Furthermore, vanillin was introduced to synthesized biobased-polyschiff vitrimer with amine as a cross-linker. The method provides cross-linked networks to obtain vitrimer with self-healing ability, hot reprocessing properties, and chemical recyclability under acid hydrolysis at 50 °C [44]. Other biobased polyamine vitrimer reported the properties of mechanical and thermal polyamine vitrimer with different cross-linking degrees, using an in-situ condensation reaction on vanillin-based aldehydes and commercial biobased aliphatic amine agents [148]. These procedures are considered a significant opportunity for designing self-healing and chemically recyclable polymer materials from lignin-based monomers. Moreover, the investigation relying on acetal exchange reactions has been demonstrated. This approach is succeeded

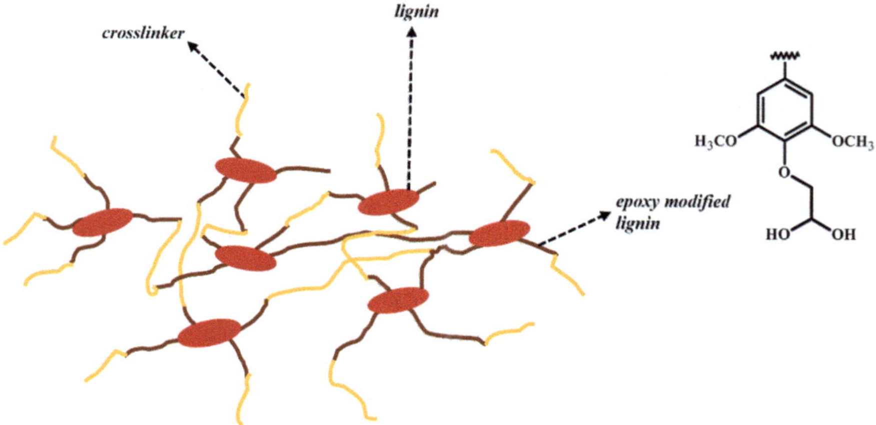

Fig. 20 Lignin-epoxy vitrimer curing mechanism

prepared a catalyst-free product from softwood Kraft lignin by exploiting aliphatic and phenolic hydroxyl groups to conduct a thermal catalysis reaction. The dynamic network can be resulted by involving poly (ethylene glycol) divinyl ether (PVD) on thermally labile acetal linkages [82].

8 Lignin-Based Adhesive Applications for Wood Bonding

The modification of lignin has shown outstanding reviews to create lignin-based adhesives for wood bonding. This utilization-based adhesive can reduce the application of synthetic resins derived from petroleum with minimum cost and environmentally friendliness. In addition, lignin-based adhesives can reduce the risk of health issues related to the formaldehyde emission from synthetic polymer-based adhesives [37]. The application of adhesives is cured by pressing with either hot or cold press under a certain temperature to build adhesion between adhesives and wood surfaces (Fig. 21). An optimal condition might depend on adhesive properties such as adhesive formulations, viscosity, pH, and reactivity. The processing parameters such as hot-pressing temperature and time affect their bonding performance between wood materials [104, 134]. In addition, the formulation or level of resin plays a crucial role in determining wood bonding. The exact formulation can optimize adhesion performance such as adhesion strength and wood failure. The application of lignin-based adhesives without formaldehyde has been done to prepare more environmentally friendly products. Particleboard without formaldehyde-based resins has fulfilled international standards with comparable internal bond strength [31]. Moreover, an advanced modification of using nanolignin as raw material was used for phenol-glyoxal-lignin adhesives to make wood bonding was reported to reduce the cost of production and to substitute with synthetic phenol [141].

LPF resin adhesives are the most modification of lignin to generate comparable adhesives in polyurethane foam, epoxy glue, particleboard, and plywood. However, the mechanical and physical properties are a huge consideration to create high-value-added products. Dry and wet shear strength of lignin-based adhesive can compete for industrial applications by fulfilling the requirement of certain standards. Moreover,

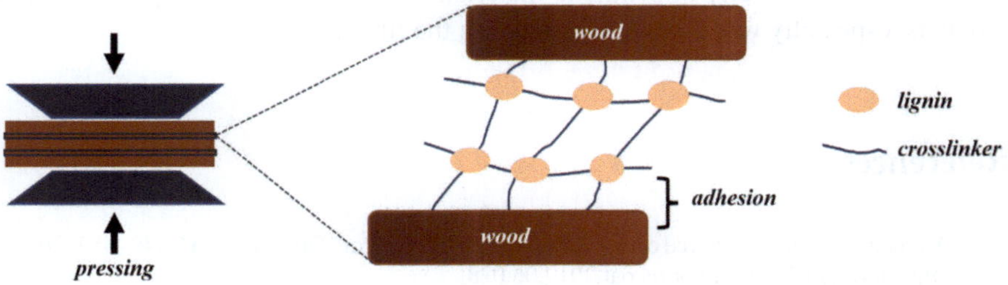

Fig. 21 Schematic diagram of applied adhesives for wood bonding

Table 4 Lignin application for wood adhesives

Sample	Dry shear strength (MPa)	Application	References
Kraft lignin	3.64	Polyurethane	Gadhave et al. [37]
Corn stover	3.40	Glue	Kalami et al. [58]
Corn stover	3.90	Glue	Siahkamari et al. [106]
Organosolv	~4.00	Particleboard	Podschun et al. [93]
Lignosulfonate	~4.30	Particleboard	Podschun et al. [93]
Organosolv	1.20	Plywood	Navarrete et al. [87]
Spent coffee	2.15	Plywood	Mennani et al. [79]

the wet shear strength could suggest the water resistance of wood adhesives [48, 134]. Numerous applications with adhesion performance could compete with industrial uses, as presented in Table 4.

9 Conclusion

Lignin has been described as one of bio-based polymers from lignocellulosic materials with tons of intended purposes. The production of suitable raw material from lignin has found many modifications of heterogenous structures of lignin for further processing. Lignin modification is mainly considered as raw material to produce lignin-based wood adhesives. It has been proven that various modifications have the capacity to alter lignin reactivity and molecular weight, generating lignin resin with comparable adhesive performance. Lignin-based adhesives have been successfully applied to make lignin-phenol formaldehyde (LPF), lignin-based epoxy resin, and lignin-based polyurethane. Moreover, novel ideas that require additional development have been found through the modification of nanoparticles and vitrimer as these approaches have the potential for a wide range of applications in the future. As a substituted material, lignin provides some advantages such as nontoxic, environmentally friendly, biodegradable, and low cost. In conclusion, the utilization of lignin as abundant aromatic bio-based polymer is expected to answer environmental issues due to the toxicity of synthetic materials for making synthetic polymer base products, especially wood-based adhesives in the future.

References

1. Ahmadi, Z.: Nanostructured epoxy adhesives: a review. Prog. Org. Coat. **135**, 449–453 (2019). https://doi.org/10.1016/j.porgcoat.2019.06.028
2. Alén, R.: Structure and chemical composition of wood. In: Forest Products Chemistry, pp. 11–57 (2000)

3. Aliakbari, M., Jazani, O.M., Sohrabian, M., Jouyandeh, M., Saeb, M.R.: Multi-nationality epoxy adhesives on trial for future nanocomposite developments. Prog. Org. Coat. **133**, 376–386 (2019). https://doi.org/10.1016/j.porgcoat.2019.04.076
4. Alonso, M.V., Oliet, M., Rodríguez, F., Astarloa, G., Echeverría, J.M.: Use of a methylolated softwood ammonium lignosulfonate as partial substitute of phenol in resol resins manufacture. J. Appl. Polym. Sci. **94**(2), 643–650 (2004). https://doi.org/10.1002/app.20887
5. Alraddadi, M.A., Chiaradia, V., Stubbs, C.J., Worch, J.C., Dove, A.P.: Renewable and recyclable covalent adaptable networks based on bio-derived lipoic acid. Polym. Chem. **12**(40), 5796–5802 (2021). https://doi.org/10.1039/d1py00754h
6. Altuna, F.I., Hoppe, C.E., Williams, R.J.J.: Epoxy vitrimers: the effect of transesterification reactions on the network structure. Polymers (Basel) **10**(1), 43 (2018). https://doi.org/10.3390/polym10010043
7. Ang, A.F., Ashaari, Z., Bakar, E.S., Ibrahim, N.A.: Possibility of enhancing the dimensional stability of jelutong (Dyera costulata) wood using glyoxalated alkali lignin-phenolic resin as bulking agent. Eur. J. Wood Wood Prod. **76**(1), 269–282 (2017). https://doi.org/10.1007/s00107-016-1139-6
8. Arefmanesh, M., Nikafshar, S., Master, E.R., Nejad, M.: From acetone fractionation to lignin-based phenolic and polyurethane resins. Ind. Crops Prod. **178**, 114604 (2022). https://doi.org/10.1016/j.indcrop.2022.114604
9. Auvergne, R., Caillol, S., David, G., Boutevin, B., Pascault, J.P.: Biobased thermosetting epoxy: present and future. Chem. Rev. **114**, 1082–1115 (2014)
10. Avelino, F., de Oliveira, D.R., Mazzetto, S.E., Lomonaco, D.: Poly(methyl methacrylate) films reinforced with coconut shell lignin fractions to enhance their UV-blocking, antioxidant and thermo-mechanical properties. Int. J. Biol. Macromol. **125**, 171–180 (2019). https://doi.org/10.1016/j.ijbiomac.2018.12.043
11. Aziz, N.A., Latip, A.F.A., Peng, L.C., Latif, N.H.A., Brosse, N., Hashim, R., Hussin, M.H.: Reinforced lignin-phenol-glyoxal (LPG) wood adhesives from coconut husk. Int. J. Biol. Macromol. **141**, 185–196 (2019). https://doi.org/10.1016/j.ijbiomac.2019.08.255
12. Belgacem, M.N., Gandini, A.: Monomers, Polymers and Composites from Renewable Resources, First edn. Elsevier (2008)
13. Benar, P., Gonqalves, A.R., Mandelli, D., Schuchardt, U.: Eucalyptus organosolv lignins: study of the hydroxymethylation and use in resols. Bioresour. Technol. **68**, 11–16 (1999)
14. Berrima, B., Mortha, G., El, A.E., Belgacem, M.N.: Oxypropylation of soda lignin: characterization and application in polyurethane foams production. Cellul. Chem. Technol. **50**(9–10), 941–950 (2016)
15. Brodin, I., Sjöholm, E., Gellerstedt, G.: Kraft lignin as feedstock for chemical products: the effects of membrane filtration. Holzforschung **63**(3), 290–297 (2009). https://doi.org/10.1515/HF.2009.049
16. Carvajal, J.C., Gómez, Á., Cardona, C.A.: Comparison of lignin extraction processes: economic and environmental assessment. Bioresour. Technol. **214**, 468–476 (2016). https://doi.org/10.1016/j.biortech.2016.04.103
17. Cassales, A., Ramos, L.A., Frollini, E.: Synthesis of bio-based polyurethanes from Kraft lignin and castor oil with simultaneous film formation. Int. J. Biol. Macromol. **145**, 28–41 (2020). https://doi.org/10.1016/j.ijbiomac.2019.12.173
18. Cateto, C.A., Barreiro, M.F., Rodrigues, A.E., Belgacem, M.N.: Optimization study of lignin oxypropylation in view of the preparation of polyurethane rigid foams. Ind. Eng. Chem. Res. **48**(5), 2583–2589 (2009). https://doi.org/10.1021/ie801251r
19. Cederholm, L., Xu, Y., Tagami, A., Sevastyanova, O., Odelius, K., Hakkarainen, M.: Microwave processing of lignin in green solvents: a high-yield process to narrow-dispersity oligomers. Ind. Crops Prod. **145**, 112152 (2020). https://doi.org/10.1016/j.indcrop.2020.112152
20. Chen, S., Xin, Y., Zhao, C.: Multispectroscopic analysis in the synthesis of lignin-based biophenolic resins. ACS Sustain. Chem. Eng. **9**(46), 15653–15660 (2021). https://doi.org/10.1021/acssuschemeng.1c06135

21. Chen, X., Pizzi, A., Zhang, B., Zhou, X., Fredon, E., Gerardin, C., Du, G.: Particleboard bio-adhesive by glyoxalated lignin and oxidized dialdehyde starch crosslinked by urea. Wood Sci. Technol. **56**(1), 63–85 (2022). https://doi.org/10.1007/s00226-021-01344-z
22. Chen, Y., Zhang, H., Zhu, Z., Fu, S.: High-value utilization of hydroxymethylated lignin in polyurethane adhesives. Int. J. Biol. Macromol. **152**, 775–785 (2020). https://doi.org/10.1016/j.ijbiomac.2020.02.321
23. Crestini, C., Lange, H., Sette, M., Argyropoulos, D.S.: On the structure of softwood Kraft lignin. Green Chem. **19**(17), 4104–4121 (2017). https://doi.org/10.1039/c7gc01812f
24. Cui, C., Sun, R., Argyropoulos, D.S.: Fractional precipitation of softwood Kraft lignin: isolation of narrow fractions common to a variety of lignins. ACS Sustain. Chem. Eng. **2**(4), 959–968 (2014). https://doi.org/10.1021/sc400545d
25. da Silva, S.H.F., Gordobil, O., Labidi, J.: Organic acids as a greener alternative for the precipitation of hardwood Kraft lignins from the industrial black liquor. Int. J. Biol. Macromol. **142**, 583–591 (2020). https://doi.org/10.1016/j.ijbiomac.2019.09.133
26. Denissen, W., Winne, J.M., Du Prez, F.E.: Vitrimers: permanent organic networks with glass-like fluidity. Chem. Sci. **7**(1), 30–38 (2016). https://doi.org/10.1039/c5sc02223a
27. Doherty, W.O.S., Mousavioun, P., Fellows, C.M.: Value-adding to cellulosic ethanol: lignin polymers. Ind. Crops Prod. **33**, 259–276 (2011)
28. Dorrestijn, E., Laarhoven, L.J.J., Arends, I.W.C.E., Mulder, P.: The occurrence and reactivity of phenoxyl linkages in lignin and low rank coal. J. Anal. Appl. Pyrolysis **54**, 153–192 (2000)
29. Duval, A., Vilaplana, F., Crestini, C., Lawoko, M.: Solvent screening for the fractionation of industrial Kraft lignin. Holzforschung **70**(1), 11–20 (2016). https://doi.org/10.1515/hf-2014-0346
30. el Hage, R., Brosse, N., Sannigrahi, P., Ragauskas, A.: Effects of process severity on the chemical structure of miscanthus ethanol organosolv lignin. Polym. Degrad. Stab. **95**(6), 997–1003 (2010). https://doi.org/10.1016/j.polymdegradstab.2010.03.012
31. El Mansouri, N.E., Pizzi, A., Salvado, J.: Lignin-based polycondensation resins for wood adhesives. J. Appl. Polym. Sci. **103**(3), 1690–1699 (2007). https://doi.org/10.1002/app.25098
32. Eraghi Kazzaz, A., Fatehi, P.: Technical lignin and its potential modification routes: a mini-review. Ind. Crops Prod. **154**, 112732 (2020). https://doi.org/10.1016/j.indcrop.2020.112732
33. Feldman, D., Banu, D., Luchian, C., Wang, J.: Epoxy-lignin polyblends: correlation between polymer interaction and curing temperature. J. Appl. Polym. Sci. **42**, 1307–1318 (1991)
34. Ferdosian, F., Yuan, Z., Anderson, M., Xu, C.: Synthesis of lignin-based epoxy resins: optimization of reaction parameters using response surface methodology. RSC Adv. **4**(60), 31745–31753 (2014). https://doi.org/10.1039/c4ra03978e
35. Foyer, G., Chanfi, B.H., Boutevin, B., Caillol, S., David, G.: New method for the synthesis of formaldehyde-free phenolic resins from lignin-based aldehyde precursors. Eur. Polym. J. **74**, 296–309 (2016). https://doi.org/10.1016/j.eurpolymj.2015.11.036
36. Fredheim, G.E., Braaten, S.M., Christensen, B.E.: Molecular weight determination of lignosulfonates by size-exclusion chromatography and multi-angle laser light scattering. J. Chromatogr. A **942**, 191–199 (2002)
37. Gadhave, R.V., Srivastava, S., Mahanwar, P.A., Gadekar, P.T.: Lignin: renewable raw material for adhesive. Open J. Polym. Chem. **09**(02), 27–38 (2019). https://doi.org/10.4236/ojpchem.2019.92003
38. Gao, S., Cheng, Z., Zhou, X., Liu, Y., Chen, R., Wang, J., Wang, C., Chu, F., Xu, F., Zhang, D.: Unexpected role of amphiphilic lignosulfonate to improve the storage stability of urea formaldehyde resin and its application as adhesives. Int. J. Biol. Macromol. **161**, 755–762 (2020). https://doi.org/10.1016/j.ijbiomac.2020.06.135
39. Gao, W., Fatehi, P.: Lignin for polymer and nanoparticle production: current status and challenges. Can. J. Chem. Eng. **97**(11), 2827–2842 (2019). https://doi.org/10.1002/cjce.23620
40. Gao, W., Inwood, J.P.W., Fatehi, P.: Sulfonation of hydroxymethylated lignin and its application. J. Bioresour. Bioprod. **4**(2), 80–88 (2019). https://doi.org/10.21967/jbb.v4i2.228
41. Gao, Z., Lang, X., Chen, S., Zhao, C.: Mini-review on the synthesis of lignin-based phenolic resin. Energy Fuels **35**, 18385–18395 (2021). https://doi.org/10.1021/acs.energyfuels.1c03177

42. García, A., Toledano, A., Serrano, L., Egüés, I., González, M., Marín, F., Labidi, J.: Characterization of lignins obtained by selective precipitation. Sep. Purif. Technol. **68**(2), 193–198 (2009). https://doi.org/10.1016/j.seppur.2009.05.001
43. Gellerstedt, G.: Softwood Kraft lignin: raw material for the future. Ind. Crops Prod. **77**, 845–854 (2015). https://doi.org/10.1016/j.indcrop.2015.09.040
44. Geng, H., Wang, Y., Yu, Q., Gu, S., Zhou, Y., Xu, W., Zhang, X., Ye, D.: Vanillin-based polyschiff vitrimers: reprocessability and chemical recyclability. ACS Sustain. Chem. Eng. **6**(11), 15463–15470 (2018). https://doi.org/10.1021/acssuschemeng.8b03925
45. Ghahri, S., Park, B.-D.: Self-crosslinking of acetone-fractionated and glyoxalated hardwood Kraft lignin as bio-adhesives for wood bonding. Ind. Crops Prod. **206**, 117711 (2023)
46. Gioia, C., Colonna, M., Tagami, A., Medina, L., Sevastyanova, O., Berglund, L.A., Lawoko, M.: Lignin-based epoxy resins: unravelling the relationship between structure and material properties. Biomacromol **21**(5), 1920–1928 (2020). https://doi.org/10.1021/acs.biomac.0c00057
47. Goncalves, A.R., Benar, P.: Hydroxymethylation and oxidation of organosolv lignins and utilization of the products. Bioresour. Technol. **79**, 103–111 (2001)
48. Gong, X., Meng, Y., Lu, J., Tao, Y., Cheng, Y., Wang, H.: A review on lignin-based phenolic resin adhesive. Macromol. Chem. Phys. **223**, 2100434 (2022). https://doi.org/10.1002/macp.202100434
49. Guadix-Montero, S., Sankar, M.: Review on catalytic cleavage of C-C inter-unit linkages in lignin model compounds: towards lignin depolymerisation. Top. Catal. **61**(3–4), 183–198 (2018). https://doi.org/10.1007/s11244-018-0909-2
50. Hattali, S., Benaboura, A., Dumarçay, S., Gérardin, P.: Evaluation of Alfa grass soda lignin as a filler for novolak molding powder. J. Appl. Polym. Sci. **97**(3), 1065–1068 (2005). https://doi.org/10.1002/app.21243
51. Henn, K.A., Forssell, S., Pietiläinen, A., Forsman, N., Smal, I., Nousiainen, P., Bangalore Ashok, R.P., Oinas, P., Österberg, M.: Interfacial catalysis and lignin nanoparticles for strong fire- and water-resistant composite adhesives. Green Chem. **24**(17), 6487–6500 (2022). https://doi.org/10.1039/d2gc01637k
52. Huang, J., Fu, S., Gan, L.: Lignin Chemistry and Applications. Elsevier, Netherland (2019)
53. Inkrod, C., Raita, M., Champreda, V., Laosiripojana, N.: Characteristics of lignin extracted from different lignocellulosic materials via organosolv fractionation. Bioenergy Res. **11**(2), 277–290 (2018). https://doi.org/10.1007/s12155-018-9895-2
54. Jawerth, M., Johansson, M., Lundmark, S., Gioia, C., Lawoko, M.: Renewable thiol-ene thermosets based on refined and selectively allylated industrial lignin. ACS Sustain. Chem. Eng. **5**(11), 10918–10925 (2017). https://doi.org/10.1021/acssuschemeng.7b02822
55. Jawerth, M., Lawoko, M., Lundmark, S., Perez-Berumen, C., Johansson, M.: Allylation of a lignin model phenol: a highly selective reaction under benign conditions towards a new thermoset resin platform. RSC Adv. **6**(98), 96281–96288 (2016). https://doi.org/10.1039/c6ra21447a
56. Jawerth, M.E., Brett, C.J., Terrier, C., Larsson, P.T., Lawoko, M., Roth, S.V., Lundmark, S., Johansson, M.: Mechanical and morphological properties of lignin-based thermosets. ACS Appl. Polym. Mater. **2**(2), 668–676 (2020). https://doi.org/10.1021/acsapm.9b01007
57. Jiang, X., Liu, J., Du, X., Hu, Z., Chang, H.M., Jameel, H.: Phenolation to improve lignin reactivity toward thermosets application. ACS Sustain. Chem. Eng. **6**(4), 5504–5512 (2018). https://doi.org/10.1021/acssuschemeng.8b00369
58. Kalami, S., Arefmanesh, M., Master, E., Nejad, M.: Replacing 100% of phenol in phenolic adhesive formulations with lignin. J. Appl. Polym. Sci. **134**(30), 45124 (2017). https://doi.org/10.1002/app.45124
59. Kalami, S., Chen, N., Borazjani, H., Nejad, M.: Comparative analysis of different lignins as phenol replacement in phenolic adhesive formulations. Ind. Crops Prod. **125**, 520–528 (2018). https://doi.org/10.1016/j.indcrop.2018.09.037
60. Karaaslan, M.A., Lin, L.T., Ko, F., Renneckar, S.: Carbon aerogels from softwood Kraft lignin for high performance supercapacitor electrodes. Front. Mater. **9**, 894061 (2022). https://doi.org/10.3389/fmats.2022.894061

61. Kimiaei, E., Farooq, M., Grande, R., Meinander, K., Österberg, M.: Lignin nanoparticles as an interfacial modulator in tough and multi-resistant cellulose–polycaprolactone nanocomposites based on a pickering emulsions strategy. Adv. Mater. Interfaces **9**(27), 2200988 (2022). https://doi.org/10.1002/admi.202200988
62. Kleinert, M., Barth, T.: Phenols from lignin. Chem. Eng. Technol. **31**(5), 736–745 (2008). https://doi.org/10.1002/ceat.200800073
63. Kloxin, C.J., Scott, T.F., Adzima, B.J., Bowman, C.N.: Covalent adaptable networks (CANs): a unique paradigm in cross-linked polymers. Macromolecules **43**(6), 2643–2653 (2010). https://doi.org/10.1021/ma902596s
64. Lee, W.J., Chang, K.C., Tseng, I.M.: Properties of phenol-formaldehyde resins prepared from phenol-liquefied lignin. J. Appl. Polym. Sci. **124**(6), 4782–4788 (2012). https://doi.org/10.1002/app.35539
65. Li, H., Deng, Y., Liu, B., Ren, Y., Liang, J., Qian, Y., Qiu, X., Li, C., Zheng, D.: Preparation of nanocapsules via the self-assembly of Kraft lignin: a totally green process with renewable resources. ACS Sustain. Chem. Eng. **4**(4), 1946–1953 (2016). https://doi.org/10.1021/acssuschemeng.5b01066
66. Li, J., Wang, W., Zhang, S., Gao, Q., Zhang, W., Li, J.: Preparation and characterization of lignin demethylated at atmospheric pressure and its application in fast curing biobased phenolic resins. RSC Adv. **6**(71), 67435–67443 (2016). https://doi.org/10.1039/c6ra11966b
67. Li, J., Zhang, J., Zhang, S., Gao, Q., Li, J., Zhang, W.: Alkali lignin depolymerization under eco-friendly and cost-effective NaOH/urea aqueous solution for fast curing bio-based phenolic resin. Ind. Crops Prod. **120**, 25–33 (2018). https://doi.org/10.1016/j.indcrop.2018.04.027
68. Liao, J.J., Latif, N.H.A., Trache, D., Brosse, N., Hussin, M.H.: Current advancement on the isolation, characterization and application of lignin. Int. J. Biol. Macromol. **162**, 985–1024 (2020). https://doi.org/10.1016/j.ijbiomac.2020.06.168
69. Lievonen, M., Valle-Delgado, J.J., Mattinen, M.L., Hult, E.L., Lintinen, K., Kostiainen, M.A., Paananen, A., Szilvay, G.R., Setälä, H., Österberg, M.: A simple process for lignin nanoparticle preparation. Green Chem. **18**(5), 1416–1422 (2016). https://doi.org/10.1039/c5gc01436k
70. Lizundia, E., Sipponen, M.H., Greca, L.G., Balakshin, M., Tardy, B.L., Rojas, O.J., Puglia, D.: Multifunctional lignin-based nanocomposites and nanohybrids. Green Chem. **23**, 6698–6760 (2021)
71. Lucherelli, M.A., Duval, A., Avérous, L.: Biobased vitrimers: towards sustainable and adaptable performing polymer materials. Prog. Polym. Sci. **127**, 101515 (2022). https://doi.org/10.1016/j.progpolymsci.2022.101515
72. Ma, X., Chen, J., Zhu, J., Yan, N.: Lignin-based polyurethane: recent advances and future perspectives. Macromol. Rapid Commun. **42**, 2000492 (2021). https://doi.org/10.1002/marc.202000492
73. Ma, Y., Zhao, X., Chen, X., Wang, Z.: An approach to improve the application of acid-insoluble lignin from rice hull in phenol-formaldehyde resin. Colloids Surf A Physicochem Eng Asp **377**(1–3), 284–289 (2011). https://doi.org/10.1016/j.colsurfa.2011.01.006
74. Mahmood, N., Yuan, Z., Schmidt, J., Xu, C.C.: Hydrolytic depolymerization of hydrolysis lignin: effects of catalysts and solvents. Bioresour. Technol. **190**, 416–419 (2015). https://doi.org/10.1016/j.biortech.2015.04.074
75. Malyar, Y.N., Kazachenko, A.S., Vasilyeva, N.Y., Fetisova, O.Y., Borovkova, V.S., Miroshnikova, A.V., Levdansky, A.V., Skripnikov, A.M.: Sulfation of wheat straw soda lignin: role of solvents and catalysts. Catal. Today **397–399**, 397–406 (2022). https://doi.org/10.1016/j.cattod.2021.07.033
76. El Mansouri, N.-E., Yuan, Q., Huang, F.: Synthesis and characterization of Kraft lignin-based epoxy resins. BioResources **6**(3), 2492–2503 (2011)
77. Mattinen, M.L., Riviere, G., Henn, A., Nugroho, R.W.N., Leskinen, T., Nivala, O., Valle-Delgado, J.J., Kostiainen, M.A., Österberg, M.: Colloidal lignin particles as adhesives for soft materials. Nanomaterials **8**(12), 1001 (2018). https://doi.org/10.3390/nano8121001
78. Meneses, D.B., Montes De Oca-Vásquez, G., Roberto Vega-Baudrit, J., Rojas-Álvarez, M., Corrales-Castillo, J., Murillo-Araya, L.C.: Pretreatment methods of lignocellulosic wastes

into value-added products: recent advances and possibilities. Biomass Convers. Biorefin. **12**, 547–564 (2022). https://doi.org/10.1007/s13399-020-00722-0/Published
79. Mennani, M., Ait Benhamou, A., Kasbaji, M., Boussetta, A., Ablouh, E.H., Kassab, Z., El Achaby, M., Boussetta, N., Grimi, N., Moubarik, A.: Insights on the physico-chemical properties of alkali lignins from different agro-industrial residues and their use in phenol-formaldehyde wood adhesive formulation. Int. J. Biol. Macromol. **221**, 149–162 (2022). https://doi.org/10.1016/j.ijbiomac.2022.08.191
80. Mhatre, S.V., Mahajan, J.S., Epps, T.H., Korley, L.S.T.J.: Lignin-derivable alternatives to petroleum-derived non-isocyanate polyurethane thermosets with enhanced toughness. Mater. Adv. **4**, 110–121 (2022). https://doi.org/10.1039/d2ma00895e
81. Montarnal, D., Capelot, M., Tournilhac, F., Leibler, L.: Silica-like malleable materials from permanent organic networks. Science (1979) **334**(6058), 965–968 (2011). https://doi.org/10.1126/science.1211649
82. Moreno, A., Morsali, M., Sipponen, M.H.: Catalyst-free synthesis of lignin vitrimers with tunable mechanical properties: circular polymers and recoverable adhesives. ACS Appl. Mater. Interfaces **13**(48), 57952–57961 (2021). https://doi.org/10.1021/acsami.1c17412
83. Mun, J.S., Pe, J.A., Mun, S.P.: Chemical characterization of Kraft lignin prepared from mixed hardwoods. Molecules **26**(16), 4861 (2021). https://doi.org/10.3390/molecules26164861
84. Nadji, H., Bruzzèse, C., Belgacem, M.N., Benaboura, A., Gandini, A.: Oxypropylation of lignins and preparation of rigid polyurethane foams from the ensuing polyols. Macromol. Mater. Eng. **290**(10), 1009–1016 (2005). https://doi.org/10.1002/mame.200500200
85. Nahra, L.R., Rezende, M.C., Oliveira, M.P., Guerrini, L.M.: Glyoxalation of Kraft lignin and optimization of electrospinning process parameters for producing polyacrylonitrile/KL nanomats for potential applications as carbon material. J. Polym. Res. **27**(11), 1–14 (2020). https://doi.org/10.1007/s10965-020-02304-0
86. Nakamura, Y., Sawada, T., Kuno, K., Nakamoto, Y.: Resinification of woody lignin and its characteristics on safety and biodegradation. J. Chem. Eng. Jpn. **34**(10), 1309–1312 (2001)
87. Navarrete, P., Mansouri, H.R., Pizzi, A., Tapin-Lingua, S., Benjelloun-Mlayah, B., Pasch, H., Rigolet, S.: Wood panel adhesives from low molecular mass lignin and tannin without synthetic resins. J. Adhes. Sci. Technol. **24**(8), 1597–1610 (2010). https://doi.org/10.1163/016942410X500972
88. Over, L.C., Meier, M.A.R.: Sustainable allylation of organosolv lignin with diallyl carbonate and detailed structural characterization of modified lignin. Green Chem. **18**(1), 197–207 (2016). https://doi.org/10.1039/c5gc01882j
89. Pang, B., Yang, S., Fang, W., Yuan, T.Q., Argyropoulos, D.S., Sun, R.C.: Structure-property relationships for technical lignins for the production of lignin-phenol-formaldehyde resins. Ind. Crops Prod. **108**, 316–326 (2017). https://doi.org/10.1016/j.indcrop.2017.07.009
90. Paysepar, H., Hu, Y., Feng, S., Yuan, Z., Shui, H., Xu, C.C.: Bio-phenol formaldehyde (BPF) resoles prepared using phenolic extracts from the biocrude oils derived from hydrothermal liquefaction of hydrolysis lignin. React. Funct. Polym. **146**, 104442 (2020). https://doi.org/10.1016/j.reactfunctpolym.2019.104442
91. Piccinino, D., Capecchi, E., Delfino, I., Crucianelli, M., Conte, N., Avitabile, D., Saladino, R.: Green and scalable preparation of colloidal suspension of lignin nanoparticles and its application in eco-friendly sunscreen formulations. ACS Omega **6**(33), 21444–21456 (2021). https://doi.org/10.1021/acsomega.1c02268
92. Pinto, J.A., Prieto, M.A., Ferreira, I.C.F.R., Belgacem, M.N., Rodrigues, A.E., Barreiro, M.F.: Analysis of the oxypropylation process of a lignocellulosic material, almond shell, using the response surface methodology (RSM). Ind. Crops Prod. **153**, 112542 (2020). https://doi.org/10.1016/j.indcrop.2020.112542
93. Podschun, J., Stücker, A., Buchholz, R.I., Heitmann, M., Schreiber, A., Saake, B., Lehnen, R.: Phenolated lignins as reactive precursors in wood veneer and particleboard adhesion. Ind. Eng. Chem. Res. **55**(18), 5231–5237 (2016). https://doi.org/10.1021/acs.iecr.6b00594
94. Qin, J., Woloctt, M., Zhang, J.: Use of polycarboxylic acid derived from partially depolymerized lignin as a curing agent for epoxy application. ACS Sustain. Chem. Eng. **2**(2), 188–193 (2014). https://doi.org/10.1021/sc400227v

95. Ribca, I., Jawerth, M.E., Brett, C.J., Lawoko, M., Schwartzkopf, M., Chumakov, A., Roth, S.V., Johansson, M.: Exploring the effects of different cross-linkers on lignin-based thermoset properties and morphologies. ACS Sustain. Chem. Eng. **9**(4), 1692–1702 (2021). https://doi.org/10.1021/acssuschemeng.0c07580
96. Rico-García, D., Ruiz-Rubio, L., Pérez-Alvarez, L., Hernández-Olmos, S.L., Guerrero-Ramírez, G.L., Vilas-Vilela, J.L.: Lignin-based hydrogels: synthesis and applications. Polymers (Basel) **12**(81), 1–23 (2020). https://doi.org/10.3390/polym12010081
97. Rodríguez, A., Moral, A., Jiménez, L.: Use of diethanolamine to obain cellulosics pulps from solid fraction of hydrothermal treatment of rice straw. AFINIDAD LXVI **539**, 20–26 (2009)
98. Sadeghifar, H., Cui, C., Argyropoulos, D.S.: Toward thermoplastic lignin polymers. Part 1. Selective masking of phenolic hydroxyl groups in Kraft lignins via methylation and oxypropylation chemistries. Ind. Eng. Chem. Res. **51**(51), 16713–16720 (2012). https://doi.org/10.1021/ie301848j
99. Sadeghifar, H., Sadeghifar, H., Ragauskas, A., Ragauskas, A., Ragauskas, A., Ragauskas, A.: Perspective on technical lignin fractionation. ACS Sustain. Chem. Eng. **8**(22), 8086–8101 (2020). https://doi.org/10.1021/acssuschemeng.0c01348
100. Saffar, T., Bouafif, H., Braghiroli, F.L., Magdouli, S., Langlois, A., Koubaa, A.: Production of bio-based polyol from oxypropylated pyrolytic lignin for rigid polyurethane foam application. Waste Biomass Valoriz. **11**(11), 6411–6427 (2020). https://doi.org/10.1007/s12649-019-00876-7
101. Saidane, D., Barbe, J.C., Birot, M., Deleuze, H.: Preparation of functionalized Kraft lignin beads. J. Appl. Polym. Sci. **116**(2), 1184–1189 (2010). https://doi.org/10.1002/app.31659
102. Saito, T., Perkins, J.H., Jackson, D.C., Trammel, N.E., Hunt, M.A., Naskar, A.K.: Development of lignin-based polyurethane thermoplastics. RSC Adv. **3**(44), 21832–21840 (2013). https://doi.org/10.1039/c3ra44794d
103. Sameni, J.K.: Physico-chemical characterization of lignin isolated from industrial sources for advanced applications. University of Toronto (2015)
104. Saražin, J., Pizzi, A., Amirou, S., Schmiedl, D., Šernek, M., Šernek Organosolv Lignin, M.: Organosolv lignin for Non-Isocyanate Based Polyurethanes (NIPU) as wood adhesive. J. Renew. Mater. Scrivener Publ. **9**(5), 881–907 (2021). https://doi.org/10.32604/jrm.2021.015047
105. Sarkar, S., Adhikari, B.: Lignin-modified phenolic resin: synthesis optimization, adhesive strength, and thermal stability. J. Adhes. Sci. Technol. **14**(9), 1179–1193 (2000). https://doi.org/10.1163/156856100743167
106. Siahkamari, M., Emmanuel, S., Hodge, D.B., Nejad, M.: Lignin-glyoxal: a fully biobased formaldehyde-free wood adhesive for interior engineered wood products. ACS Sustain. Chem. Eng. **10**(11), 3430–3441 (2022). https://doi.org/10.1021/acssuschemeng.1c06843
107. Singh, S.K., Dhepe, P.L.: Isolation of lignin by organosolv process from different varieties of rice husk: understanding their physical and chemical properties. Bioresour. Technol. **221**, 310–317 (2016). https://doi.org/10.1016/j.biortech.2016.09.042
108. Smallenburg, F., Leibler, L., Sciortino, F.: Patchy particle model for vitrimers. Phys. Rev. Lett. **111**(18), 188002 (2013). https://doi.org/10.1103/PhysRevLett.111.188002
109. Song, Y., Wang, Z., Yan, N., Zhang, R., Li, J.: Demethylation of wheat straw alkali lignin for application in phenol formaldehyde adhesives. Polymers (Basel) **8**(6), 209 (2016). https://doi.org/10.3390/polym8060209
110. Sreejaya, M.M., Sankar, R.J., Ramanunni, K., Pillai, N.P., Ramkumar, K., Anuvinda, P., Meenakshi, V.S., Sadanandan, S.: Lignin-based organic coatings and their applications: a review. Mater. Today Proc. **60**, 494–501 (2022). https://doi.org/10.1016/j.matpr.2022.01.325
111. Tang, Q., Qian, Y., Yang, D., Qiu, X., Qin, Y., Zhou, M.: Lignin-based nanoparticles: a review on their preparations and applications. Polymers (Basel) **12**(2471), 1–22 (2020). https://doi.org/10.3390/polym12112471
112. Tang, Q., Zhou, M., Li, Y., Qiu, X., Yang, D.: Formation of uniform colloidal spheres based on lignosulfonate, a renewable biomass resource recovered from pulping spent liquor. ACS Sustain. Chem. Eng. **6**(1), 1379–1386 (2018). https://doi.org/10.1021/acssuschemeng.7b03756

113. Tang, R., Xue, B., Tan, J., Guan, Y., Wen, J., Li, X., Zhao, W.: Regulating lignin-based epoxy vitrimer performance by fine-tuning the lignin structure. ACS Appl. Polym. Mater. **4**(2), 1117–1125 (2022). https://doi.org/10.1021/acsapm.1c01541
114. Tejado, A., Peña, C., Labidi, J., Echeverria, J.M., Mondragon, I.: Physico-chemical characterization of lignins from different sources for use in phenol-formaldehyde resin synthesis. Bioresour. Technol. **98**(8), 1655–1663 (2007). https://doi.org/10.1016/j.biortech.2006.05.042
115. Thébault, M., Kutuzova, L., Jury, S., Eicher, I., Zikulnig-Rusch, E.M., Kandelbauer, A.: Effect of phenolation, lignin-type and degree of substitution on the properties of lignin-modified phenol-formaldehyde impregnation resins: molecular weight distribution, wetting behavior, rheological properties and thermal curing profiles. J. Renew. Mater. **8**(6), 603–630 (2020). https://doi.org/10.32604/jrm.2020.09616
116. Tian, D., Hu, J., Bao, J., Chandra, R.P., Saddler, J.N., Lu, C.: Lignin valorization: lignin nanoparticles as high-value bio-additive for multifunctional nanocomposites. Biotechnol. Biofuels **10**(1), 1–11 (2017). https://doi.org/10.1186/s13068-017-0876-z
117. Upton, B.M., Kasko, A.M.: Strategies for the conversion of lignin to high-value polymeric materials: review and perspective. Chem. Rev. **116**, 2275–2306 (2016). https://doi.org/10.1021/acs.chemrev.5b00345
118. Van de Velde, N., Saška, J., Sever, T., Štular, D., Šobak, M., Štirn, Ž, Likozar, B., Jerman, I.: Bio-based epoxy adhesives with lignin-based aromatic monophenols replacing bisphenol A. Polymers (Basel) **13**(22), 3879 (2021). https://doi.org/10.3390/polym13223879
119. Venkatesagowda, B., Dekker, R.F.H.: Enzymatic demethylation of Kraft lignin for lignin-based phenol-formaldehyde resin applications. Biomass Convers. Biorefin. **10**, 203–225 (2020). https://doi.org/10.1007/s13399-019-00407-3/Published
120. Vishtal, A., Kraslawski, A.: Challenges in industrial applications of technical lignins. BioResources **6**(3), 3547–3568 (2011)
121. Wang, M., Zhao, Y., Li, J.: Demethylation and other modifications of industrial softwood Kraft lignin by laccase-mediators. Holzforschung **72**(5), 357–365 (2018). https://doi.org/10.1515/hf-2017-0096
122. Wang, Y.Y., Wyman, C.E., Cai, C.M., Ragauskas, A.J.: Lignin-based polyurethanes from unmodified Kraft lignin fractionated by sequential precipitation. ACS Appl. Polym. Mater. **1**(7), 1672–1679 (2019). https://doi.org/10.1021/acsapm.9b00228
123. Wang, Z., Yuan, L., Tang, C.: Sustainable elastomers from renewable biomass. Acc. Chem. Res. **50**(7), 1762–1773 (2017). https://doi.org/10.1021/acs.accounts.7b00209
124. Watumlawar, E.C., Park, B.D.: Effects of precipitation ph of black liquor on characteristics of precipitated and acetone-fractionated Kraft lignin. J. Korean Wood Sci. Technol. **51**(1), 38–48 (2023). https://doi.org/10.5658/WOOD.2023.51.1.38
125. Watumlawar, E.C., Park, B.D.: Synthesis of acetone-fractionated hardwood Kraft lignin-based adhesive crosslinked with epichlorohydrin. J. Adhes. Sci. Technol. **38**(3), 442–457 (2024). https://doi.org/10.1080/01694243.2023.2236400
126. Wibowo, E.S., Park, B.D.: Chemical and thermal characteristics of ion-exchanged lignosulfonate. Molecules **28**(6), 2755 (2023). https://doi.org/10.3390/molecules28062755
127. Wibowo, E.S., Park, B.D.: Probing the relationship between chemical structure and thermal degradation behavior of acetone fractionated Kraft lignin. J. Anal. Appl. Pyrolysis **172**, 106028 (2023). https://doi.org/10.1016/j.jaap.2023.106028
128. Wibowo, E.S., Park, B.D.: The role of acetone-fractionated Kraft lignin molecular structure on surface adhesion to formaldehyde-based resins. Int. J. Biol. Macromol. **225**, 1449–1461 (2023). https://doi.org/10.1016/j.ijbiomac.2022.11.202
129. Wu, J., Yu, X., Zhang, H., Guo, J., Hu, J., Li, M.-H.: Fully biobased vitrimers from glycyrrhizic acid and soybean oil for self-healing, shape memory, weldable, and recyclable materials. ACS Sustain. Chem. Eng. **8**(16), 6479–6487 (2020). https://doi.org/10.1021/acssuschemeng.0c01047
130. Wu, M., Peng, J., Dong, Y., Pang, J., Zhang, X.: Extraction and oxypropylation of lignin by an efficient and mild integration process from agricultural waste. Ind. Crops Prod. **172**, 114013 (2021). https://doi.org/10.1016/j.indcrop.2021.114013

131. Xu, Y.Z., Fu, P., Dai, S.L., Zhang, H.B., Bi, L.W., Jiang, J.X., Chen, Y.X.: Catalyst-free self-healing fully bio-based vitrimers derived from tung oil: Strong mechanical properties, shape memory, and recyclability. Ind. Crops Prod. **171**, 113978 (2021). https://doi.org/10.1016/j.indcrop.2021.113978
132. Xue, B., Tang, R., Xue, D., Guan, Y., Sun, Y., Zhao, W., Tan, J., Li, X.: Sustainable alternative for bisphenol A epoxy resin high-performance and recyclable lignin-based epoxy vitrimers. Ind. Crops Prod. **168**, 113583 (2021). https://doi.org/10.1016/j.indcrop.2021.113583
133. Yadav, V., Kumar, A., Bilal, M., Nguyen, T.A., Iqbal, H.M.N.: Lignin removal from pulp and paper industry waste streams and its application. In: Nanotechnology in Paper and Wood Engineering: Fundamentals, Challenges and Applications. Elsevier, pp. 265–283 (2022)
134. Yang, G., Gong, Z., Luo, X., Chen, L., Shuai, L.: Bonding wood with uncondensed lignins as adhesives. Nature **621**(7979), 511–515 (2023). https://doi.org/10.1038/s41586-023-06507-5
135. Yang, S., Wen, J.L., Yuan, T.Q., Sun, R.C.: Characterization and phenolation of biorefinery technical lignins for lignin-phenol-formaldehyde resin adhesive synthesis. RSC Adv. **4**(101), 57996–58004 (2014). https://doi.org/10.1039/c4ra09595b
136. Yang, X., Li, Z., Li, L., Li, N., Jing, F., Hu, L., Shang, Q., Zhang, X., Zhou, Y., Pan, X.: Depolymerization and demethylation of Kraft lignin in molten salt hydrate and applications as an antioxidant and metal ion scavenger. J. Agric. Food Chem. **69**(45), 13568–13577 (2021). https://doi.org/10.1021/acs.jafc.1c05759
137. Yin, Q., Yang, W., Sun, C., Di, M.: Preparation and properties of lignin-epoxy resin composite. BioResources **7**(4), 5737–5748 (2012)
138. Younesi-Kordkheili, H.: Maleated lignin coreaction with phenol-formaldehyde resins for improved wood adhesives performance. Int. J. Adhes. Adhes. **113**, 103080 (2022). https://doi.org/10.1016/j.ijadhadh.2021.103080
139. Younesi-Kordkheili, H., Pizzi, A.: Improving the properties of urea-lignin-glyoxal resin as a wood adhesive by small addition of epoxy. Int. J. Adhes. Adhes. **102**, 102681 (2020). https://doi.org/10.1016/j.ijadhadh.2020.102681
140. Younesi-Kordkheili, H., Pizzi, A.: A comparison among lignin modification methods on the properties of lignin–phenol–formaldehyde resin as wood adhesive. Polymers (Basel) **13**(20), 3502 (2021). https://doi.org/10.3390/polym13203502
141. Younesi-Kordkheili, H., Pizzi, A.: Improving the properties of phenol-lignin-glyoxal as a wood adhesive by lignin nanoparticles. Eur. J. Wood Wood Prod. **81**, 507–512 (2023). https://doi.org/10.1007/s00107-022-01904-5
142. Yu, L., Seabright, K., Bajaj, I., Keffer, D.J., Alonso, D.M., te Hsieh, C., Li, M., Chen, H., Dai, S., Gandomi, Y.A., Maravelias, C.T., Harper, D.P.: Performance and economic analysis of organosolv softwood and herbaceous lignins to activated carbons as electrode materials in supercapacitors. Front. Energy Res. **10**, 849949 (2022). https://doi.org/10.3389/fenrg.2022.849949
143. Yue, L., Guo, H., Kennedy, A., Patel, A., Gong, X., Ju, T., Gray, T., Manas-Zloczower, I.: Vitrimerization: converting thermoset polymers into vitrimers. ACS Macro Lett. **9**(6), 836–842 (2020). https://doi.org/10.1021/acsmacrolett.0c00299
144. Yuen, A.Y., Bossion, A., Veloso, A., Mecerreyes, D., Hedrick, J.L., Dove, A.P., Sardon, H.: Efficient polymerization and post-modification of: N -substituted eight-membered cyclic carbonates containing allyl groups. Polym. Chem. **9**(18), 2458–2467 (2018). https://doi.org/10.1039/c8py00231b
145. Zhang, N., Li, Z., Xiao, Y., Pan, Z., Jia, P., Feng, G., Bao, C., Zhou, Y., Hua, L.: Lignin-based phenolic resin modified with whisker silicon and its application. J. Bioresour. Bioprod. **5**(1), 67–77 (2020). https://doi.org/10.1016/j.jobab.2020.03.008
146. Zhang, T., Xue, B., Yan, Q., Yuan, Y., Tan, J., Guan, Y., Wen, J., Li, X., Zhao, W.: New kinds of non isocyanate polyurethane hybrid polymers. Ind. Crops Prod. **199**, 116706 (2023)
147. Zhong, X., Qian, Y., Huang, J., Yang, D., Deng, Y., Qiu, X.: Fabrication of lignosulfonate vesicular reverse micelles to immobilize horseradish peroxidase. Ind. Eng. Chem. Res. **55**(10), 2731–2737 (2016). https://doi.org/10.1021/acs.iecr.5b04939

148. Zhou, Z., Su, X., Liu, J., Liu, R.: Synthesis of vanillin-based polyimine vitrimers with excellent reprocessability, fast chemical degradability, and adhesion. ACS Appl. Polym. Mater. **2**(12), 5716–5725 (2020). https://doi.org/10.1021/acsapm.0c01008
149. Zoia, L., Salanti, A., Frigerio, P., Orlandi, M.: Exploring allylation and claisen rearrangement as a novel chemical modification of lignin. BioResources **9**(4), 6540–6561 (2014)
150. Zou, T., Sipponen, M.H., Henn, A., Österberg, M.: Solvent-resistant lignin-epoxy hybrid nanoparticles for covalent surface modification and high-strength particulate adhesives. ACS Nano **15**(3), 4811–4823 (2021). https://doi.org/10.1021/acsnano.0c09500
151. Zuo, L., Yao, S., Wang, W., Duan, W.: An efficient method for demethylation of aryl methyl ethers. Tetrahedron Lett. **49**(25), 4054–4056 (2008). https://doi.org/10.1016/j.tetlet.2008.04.070

Binderless Lignocellulosic Composites for Green Material

Juliana Abdul Halip, Aisyah Humaira Alias, and Balkis Fatomer A. Bakar

Abstract Binderless composites have earned significant attention among researchers as sustainable and environmentally friendly materials to conventional composites. Binderless composites do not rely on synthetic binders and utilize the natural wood elements namely fibers, particles, and veneers with inherent self-bonding properties. This paper reviews the research reported on binderless composites which are limited to binderless fibreboard, particleboard, and plywood. The focused topics include types of plant categories and binderless properties by types of composites. Most binderless fibreboards are made from wood, agricultural waste, crops, and grass, while binderless particleboards are made from agricultural waste and crops, whereas binderless plywood is made from wood species and oil palm trunks. In addition, the issues and challenges of binderless composites are discussed.

Keywords Binderless · Lignocellulosic · Composite · Fibreboard · Particleboard · Plywood

J. A. Halip (✉)
Faculty of Engineering Technology, Universiti Tun Hussein Onn Malaysia (UTHM) Pagoh Campus, Panchor, Johor, Malaysia
e-mail: julianaah@uthm.edu.my

A. H. Alias
Advanced Engineering Materials and Composites Research Centre (AEMC), Department of Mechanical and Manufacturing Engineering, Universiti Putra Malaysia, Serdang, Selangor, Malaysia

B. F. A. Bakar
Faculty of Forestry and Environment, Universiti Putra Malaysia, Serdang, Selangor, Malaysia

J. A. Halip · B. F. A. Bakar
Institute of Tropical Forestry and Forest Products (INTROP), Universiti Putra Malaysia, Serdang, Selangor, Malaysia

S. S. Osman Al Edrus et al. (eds.), *Green Lignocellulosic-Based Panels*, Engineering Materials, https://doi.org/10.1007/978-981-96-3404-0_3

1 Introduction

Nowadays, green composites gain the utmost attraction from researchers especially to ensure the sustainable use of natural plants (lignocellulosic materials) in the future. Green composites can be defined in various definitions; (i) Green or fully biodegradable composite where the composite is made of natural fibres and biodegradable resin/polymer matrix, (ii) Partially green/biodegradable composite composed of natural fibres (biofibres) and traditional petroleum-based non-biodegradable resin/polymer matrix, and (iii) Hybrid green/biodegradable composite that consisted of two or more different natural fibres in combination with biodegradable or petroleum-based resin/polymer [25]. Most of the previous studies on green composites are categorized as partially green composites or hybrid green composites. While most of the commercial composites are made from lignocellulosic materials, the bonding agents are yet using traditional petroleum-based resins namely urea formaldehyde and phenol formaldehyde which led to health concerns on the indoor air quality caused by formaldehyde emission.

To reduce this concern, manufacturing of fully green composite such as binderless or self-bonding composite is one of the environmentally friendly solutions. As stated by [41], binderless composites are environmentally friendly i.e., no formaldehyde emission, simple manufacturing process and low cost. These binderless composites could be beneficial to United Nations Sustainable Development Goals (SDGs) 3, which focus on ensuring healthy lives and promoting well-being for all at all ages initiative. Studies on binderless composites have been initiated with wood materials due to their abundancy. Apart from wood, there are many studies using non-wood materials as binderless composite including agricultural wastes (oil palm residues, coconut coir, bagasse, wheat straw, banana stem), crop (bamboo, kenaf, flax, hemp) and grass [2–4, 6, 14, 19, 24, 34, 35, 37, 39, 49, 65, 68, 74–76].

Binderless composite can be applied mostly in fibreboard, particleboard, and plywood manufacturing. During the 1970s, the possibility of bonding wood without the use of synthetic adhesives was discovered through the treatment of wood with oxidants, with lignin being identified as the primary contributor to this process [28]. Binderless composite offers some vital perceptions on the possibility of softening low molecular weight components namely lignin polymers, hemicellulose and non-crystalline cellulose at a suitable temperature to gain the plasticized matrix, which will adhere the materials together and modify the mechanical properties of self-bonding composite [2]. Lignin present in the lignocellulosic material acts as a substitution of synthetic adhesive to bind the materials. Lignin plays an important role in binderless or self-bonding composites which act as a natural binder for fibres with glass transition values in a dry state at 200 °C [21, 23, 45].

The overall manufacturing process of binderless composite is quite similar to the conventional process, starting with raw material procurement until hot pressing. Commonly, the manufacturing process of fibreboard basically occurs in wet or dry processes, while particleboard and plywood occur in the dry process. In most cases, the raw materials namely fibres, particles and veneers are produced using refiner, hammermill/flaker, and rotary lathe machines, respectively. In the next step, the raw materials, namely fibres and particles are dried and placed in a mould/forming box (lab scale) without addition of resin. Meanwhile, veneers are dried and laid up perpendicular to each other prior to the pressing process. Then, the process followed by the removal of the mould/forming box (for fibreboard and particleboard), pre-pressed, and pressed using the hot-press machine. Hot-press machine or known as heat treatment applies heat and pressure to the mat and identified as the simplest and most common method in binderless manufacturing. When heat is applied, heat activates the chemical components of the raw materials and the combination of both heat and pressure allows the melting of lignin through the entire board and provide a good distribution of lignin between the fibres [36, 78]. Meanwhile, cellulose and hemicellulose are partially hydrolyzed to simple soluble sugars that contribute to self-bonding [61].

Figure 1 illustrates the manufacturing process of binderless composites (fibreboard, particleboard and plywood), which starts with raw material and preparation of elements either to produce fibres (fibreboard), or particles (particleboard), or veneer (plywood). As compared to conventional composites, the manufacturing process of binderless composite requires no synthetic resin for blending/gluing process (Fig. 1) and relies on the self-bonding mechanism. This indicates that there is no presence of synthetic resin or adhesives during manufacturing and makes binderless composites more environmentally friendly compared to conventional composites. Apart from that, the pre-treatment process is optional for binderless composite, and it can be done either prior to the drying or during the hot-pressing processes.

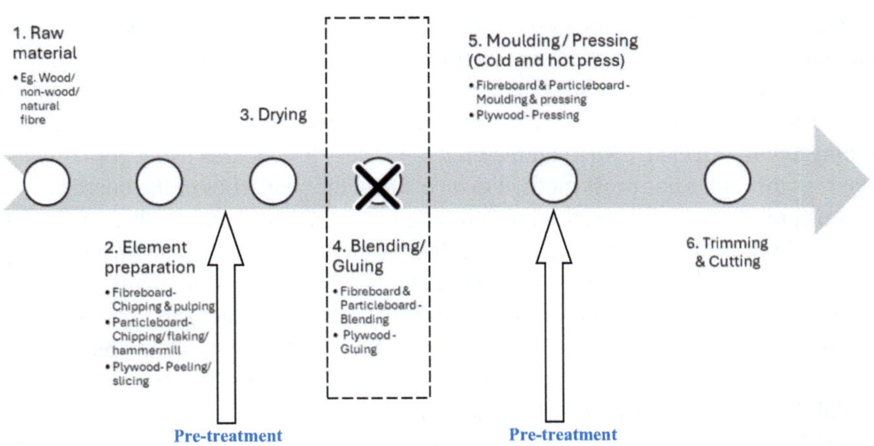

Fig. 1 Manufacturing process of conventional and binderless composites (fibreboard, particleboard and plywood)

2 Binderless Fibreboard

Fibreboard is a type of natural fibre composite that is commonly used in many applications and is made from wood or non-wood fibre components. Fibreboard is an engineered wood product made from wood pulp or fibre that is often bonded with a synthetic resin or other appropriate binders before being extruded and pressed. Formaldehyde-based resin is the most widely used resin in fibreboard because of its favourable performance and inexpensive price [7, 59]. Nonetheless, the sector is gravely concerned about the environmental issues caused by formaldehyde [1]. Thus, many studies have been conducted to shift from the above-mentioned materials to green fibreboard through the development of biobased resins, free-formaldehyde resin and binderless board [61, 64]. In the case of binderless fibreboard or self-bonding fibreboards, lignin from the fibre can be utilized as a natural resin in the manufacture of fibreboard. The main processes of self-bonding mechanism involved are lignin-lignin and lignin-polysaccharides cross-linking reactions that occur at high temperatures [41, 77]. Table 1 shows the summary of the binderless fibreboard from various sources of lignocellulosic material.

Wood is a lignocellulosic material made up of main polymeric constituents of cellulose, hemicelluloses, and lignin, as well as additional components including pectin, waxes, and fats. Generally, wood fibres were utilized in the production of fibreboard employed with urea formaldehyde (UF), melamine urea formaldehyde (MUF), phenol formaldehyde (PF) and urea melamine phenol formaldehyde (UMPF) [20, 32]. Aside from formaldehyde-based resin, self-bonded resin can also be employed in the manufacturing of fibreboard from wood due to its self-contained natural resin. For instance, [40] prepared binderless fibreboard from rubberwood fibres with different amounts of laccase enzymes at varying time intervals. They found that the lignin was converted into a stable phenoxy radical by the laccase enzyme, which also served as a suitable surface for the attachment of other components and a possible cross-linking agent. Wu et al. [70] manufactured binderless fibreboards using poplar wood shavings through a hot-press method to investigate the properties of the boards using various pre-treatment by employing cellulase enzyme. They came to the conclusion that the hydrolysis of cellulose by cellulase, which increases the hydroxyl group content, resulted in the production of high-strength binderless fibreboard.

One of the lignocellulosic resources utilized in the manufacturing of binderless fibreboard to address the wood shortage and fully utilize the natural resource is agricultural waste, such as oil palm, bagasse, rice straw, and wheat straw. Due to its availability, sustainability, and chemical content, oil palm is an ideal natural fibre for the production of value-added and environmentally friendly binderless boards [5, 21–23]. Mejía et al. [37] manufactured binderless fibreboards from steam-exploded and Fenton reagent oxidized fibres of oil palm empty fruit bunch (EFB). The physical and mechanical strengths of the binderless fibreboards produced from steam explosion-treated fibre met the board standard requirements (Colombian Standards NTC 2261). The steam explosion was found to improve the properties of the panel due to lignin

Table 1 Studies on lignocellulosic binderless fibreboard composites

Category	Fibre type	References
Wood	Softwood and hardwood fibre	Bouajila et al. [8]
	Black spruce bark	Gao et al. [16]
	Poplar	Xie et al. [71]
	Rubberwood	Nasir et al. [40]
	Poplar	Wu et al. [70]
Agricultural waste	Oil palm empty fruit bunch (EFB)	Mejia et al. [37]
	Corn	Theng et al. [62]
	Banana	Quintana et al. [50]
	Coconut coir	Araujo Junior et al. [3]
		Puspaningrum et al. [49]
		Panyakaew and Fotios [47]
	Baggase	Widyorini et al. [68]
		Panyakaew and Fotios [47]
		Luthfi et al. [34]
	Wheat straw	Yang et al. [76]
		Wang et al. [66]
		Wang et al. [67]
		Domínguez-Robles et al. et al. [13]
		Espinosa et al. [14]
Crop	Kenaf	Okuda and Sato [45]
		Widyorini et al. [69]
		Xu et al. [75]
	Bamboo	Bahari et al. [4]
		Luo et al. [33]
		Shi et al. [58]
Grass	Elephant grass	Velasquez et al. [63]
	Arundo rhizomatous grass	Vitrone et al. [65]

plasticization and redistribution onto cellulose, which eventually formed a strong fibre-to-fibre bonding by the lignin within the fibre matrix [31].

Theng et al. [62] conducted an experiment using corn thermomechanical fibres reinforced with cellulose nanofibres (CNF) to make binderless fibreboards. The binderless fibreboards produced at 0.5 wt% of CNF showed comparable mechanical properties to commercial high-density fibreboard (HDF) containing synthetic adhesives. When compared to products created just using corn stalk fibre as well as commercial HDF, it was discovered that CNF increased the resistance of the binderless fibreboards due to the good interaction between corn fibres and CNF. By uniformly distributing the fibres, the CNF and corn fibre are bonded to each other as well as to the macrofibres. Also, the high specific surface and inherent mechanical characteristics of CNF increase the number of possible hydrogen bonds and decrease the void spaces between fibres that result from tension forces and CNF shrinking [12].

Quintana et al. [50] used a steam explosion with a thermomechanical aqueous vapor process in a batch reactor to make non-synthetic binders fibreboards from banana bunch. In this study, an evaluation was done on how steam explosion pre-treatment and pressure conditions affected the physical–mechanical reactions of the fibreboard. The reduction in xylose concentration, due to an increase in pre-treatment severity from hemicelluloses hydrolysis was found to improve the dimensional stability of the fibreboards [63]. Quitana et al. [50] also mentioned that high pressing temperatures enhance the lignin's fluidity on fibres, enhancing the distribution of lignin in the board and strengthening the connections between the fibres.

Coconut coir is a waste product from coconut oil production that is taken from the coconut's outer husk and composed of 30% fibre and 70% pith, with high lignin and phenolic content [11]. Araújo Junior et al. [3] used coconut husks as a raw material for binderless fibreboards, with the main objective was to reducing the dependency on utilizing hazardous formaldehyde resin as well as investigating the potential of producing green binderless fibreboard. The findings demonstrate that pressing temperature has a significant impact on the properties of the panel. The utilization of high temperatures resulted in the creation of fibreboards displaying an improved water resistance, in compliance with the standard procedure ABNT NBR 15316-2: 2015. A study on using coconut husk to produce low density thermal insulation boards was performed by [47]. The results showed that the panels constructed from bagasse with a 350 kg/m^3 density produced at a hot press temperature of 200 °C for 13 min have better mechanical properties and satisfy JIS A 5905: 2003 standards for insulating fibreboards. The findings of the experiment also showed that increasing board density, pressing time, and temperature enhanced the mechanical qualities of the panels. A separate study investigated the effect of oxidation treatment, additional paraffin, and pressing time on the physical and mechanical properties of the binderless fibreboards from coconut fibre were analysed [49]. The fibreboards treated with a combination of oxidation, paraffin of 0.5% and 20 min of pressing time were found suitable for interior application. Most of the physical properties of the panels were found to meet the standard but not the mechanical properties. The ineffective interaction of hydroxyl radicals and lignin and weak chemical bond formed were identified as the main cause of these phenomena.

Bagasse is a waste material produced during the processing of sugarcane and is high in cellulose. Panyakaew and Fotios [47] manufactured binderless fibreboard from bagasse fibre for thermal insulation applications and compared with a coir binderless fibreboard. Binderless bagasse insulation boards produced results that were superior to coconut husk boards in terms of characteristics. The findings demonstrated that the binderless bagasse insulating boards with a density of 350 kg/m^3 satisfied the requirements except for the thickness swelling properties. Luthfi et al. [34] engineered baggase binderless fibreboad, where the drying temperatures were varied between 110 and 190 °C, and the effect of temperatures on the physical performance was investigated. The findings show that drying temperature, particularly at temperatures of 130 °C, has an inversely proportional interaction with water absorption and dimension swelling. In general, trends in physical properties like diffusion, dimension swelling, and water absorption vary due to modifications in the lignocellulosic polymer structures, which are primarily discernible at temperatures of about 150 °C.

Wheat straw is a lignocellulosic by-product of wheat cultivation that may be used to make fibreboard. Yang et al. [76] developed binderless fibreboards from wheat straw fibres and studied the effects of different laccase dosages namely 20, 40 and 60 U/g and pre-treatment conditions such as treatment time and temperature on the mechanical properties of the boards. It was discovered that the ideal conditions for excellent binderless fibreboard qualities are a treatment duration of 120 min at a temperature of 50 °C and a laccase dose of 40 U/g. When laccase is added, the fibre surface becomes more active, producing more free radicals that act on the cross-linking reaction sites. Following these reactions, the molecules of partly repolymerized active lignin interacted with one another and helped the fibre networking. Wang et al. [66] also produced binderless fibreboards from wheat straw with the addition of pulping wastewater, i.e., black liquor. The studies indicated that proteins and lignin in the black liquor assist generate new covalent bands under thermal-pressing such that have improved panel mechanical performances. In the next study, [67] investigated the difference among boards under variable hot-pressing temperatures to improve straw-based binderless fibreboard by adding pulping effluent. According to the results, straw-based fibreboards with pressing temperatures below 190 °C had higher strengths and water resistance qualities. The findings showed that, as shown in Fig. 2, the surface colour of boards darkened when the temperature was increased over 190 °C. Moreover, Fourier transform infrared (FT-IR) studies revealed that greater pressing temperatures resulted in the production of additional low-molecular components and hydrogen bonds.

The effect of different refining processes, enzymatic refining and mechanical refining, on the properties of high-density binderless fibreboards by means of the wet process was performed by [13]. Fibreboards developed from enzyme-treated fibres had noticeably better mechanical properties than those created from pulp which was not enzyme-treated. Because enzymatically treated fibres contain a relatively high percentage of highly microfibrillated fibres, resulting in a denser material with fewer voids, fibreboards made with enzymatically treated fibres demonstrated excellent bonding strength. By expanding the intensity of mechanical refining, which results in a reduction in both width and length and an increase in fines in length, mechanical refining also enhanced all the mechanical performances of the resultant fibreboards. In a different study, [14] also utilized enzymatic treatment and lignocellulose nanofibres (LCNFs) addition in the production of binderless HDF made from wheat straw. It is feasible to lower the grammage of the manufactured fibreboards, as was discovered by combining the two processes to achieve a substantial improvement.

The high yielding fibre crop kenaf may be used as a raw material in a variety of industrial processes such as particleboard, fibreboard, chipboard and insulation board. Many studies have been conducted on kenaf to create boards without binders utilizing the heat and steam methods [43–46, 69, 70, 73, 74]. A refining process

Fig. 2 Surface of binderless wheat straw fibreboards produced at different temperatures Reproduced with permission [67] [Creative Commons Attribution (CC BY) license]

was used to develop binderless fibreboards with kenaf core by [75] in order to study the boards' properties under various production settings. After applying high steam pressure and a longer cooking period, the generated boards showed excellent bonding and dimensional stability features. The boards had strong internal bonds, but because of their porous nature, they had low strength values.

Bamboo fibre is well-known for having high levels of cellulose and lignin content, as well as quick growth, high productivity, and low ash content, making it perfect for making fibreboard [51]. Using the process of digestion, [4] created binderless bamboo fibreboards. The density and lignin concentration showed a good link with the bending strength board. Another researcher [33] reported that most of the fibreboard properties binderless panels made from steam explosion pre-treatment of bamboo fibre met the requirement of the Japanese Industrial Standard (JIS). The pre-treatment was a factor in the hemicellulose breakdown products and the lignin liberation from the fibres during the steam explosion, which helped to establish the bonds in the binderless boards. Most recently, [58] investigated the absorption performance and mechanical properties of bamboo fibres according to different fibre sizes for the fabrication of special-shaped bamboo binderless fibreboards.

Researchers have been studying various ways of using different kinds of lignocellulosic residual material including grass fibre for fibreboard production. Elephant grass (*Miscanthus sinensis*) fibre was also utilized as a raw material for binderless fibreboard production. In binderless fibreboard fabrication, processing procedures are the major factors that determine the self-bonding capacity between the fibres. Velasquez et al. [63] studied the effect of the pre-treatment temperature and time, and the pressing temperature and time, on the physic mechanical responses of the fibreboard from elephant grass fibre. Low temperatures, long pre-treatment times, and high pressing temperatures are where boards exhibit their superior qualities. The results demonstrated that the pre-treatment had a significant impact on the concentration of lignin, cellulose, and hemicelluloses. The improvement in the boards' dimensional stability is directly correlated with the decline in hemicelluloses. Another research based on the steam-exploded pre-treatment of grass fibre to produce binderless fibreboard was conducted by [65]. *Arundo donax* L or perennial rhizomatous grass was used to study the effect of thickness and final heat treatment (FHT) on the physico-mechanical properties of binderless fibreboard. The findings demonstrated that the FHT had a significant impact on dimensional stability but had no impact on mechanical performance. In contrast, the thickness had no effect on physical properties, but it did cause a little shift in mechanical properties. In summary, the most recent studies on binderless fibreboard are made from agricultural waste fibres due to their availability. Additionally, most of binderless fibreboard undergo pre-treatment namely steam explosion pre-treatment and has a significant impact on their properties.

3 Binderless Particleboard

Particleboard is one of the wood-based composites that is made from lignocellulosic particles and binds together mostly using urea formaldehyde and phenol formaldehyde resins. It has been used for decades mainly for non-structural purposes, and the technology was developed to optimize the use of mill residues, sawdust and recycled wood. It can be manufactured manually (lab scale) and using layering system (pilot scale), where the inner layer consists of large particles and the outer layers are finer particles. Binderless particleboard refers to self-bonding particleboard or the manufacturing of particleboard without synthetic resin. Like fibreboard, lignin in the lignocellulosic particles can be utilized as a natural resin in the manufacture of particleboard. Table 2 tabulates the previous studies on lignocellulosic binderless particleboard composites. From observation, most current studies on binderless particleboards are made from agricultural wastes and crops.

In the manufacturing of binderless composites, the pressing times and pressing temperatures are generally higher compared to conventional composites. Mostly, increasing pressing time and temperature reduces the water absorption and improves the dimension stability of binderless composite. In a prior study, three pressing temperatures used in a binderless oil palm trunk (OPT) particleboard are 160, 180 and 200 °C [21, 23]. Their findings reported that the water absorption (WA) and thickness swelling (TS) of the boards decreased with increasing press temperature. The board pressed at 200 °C has the highest modulus of rupture (MOR) compared to others with values of 5.73 MPa. However, conventional heat treatment binderless composite has poor dimensional stability. Therefore, pretreatment processes such as steam pretreatment, chemical pretreatments, enzyme pretreatments, preheating,

Table 2 Studies on lignocellulosic binderless particleboard composites

Category	Type	References
Agricultural waste	Oil palm trunk	Hashim et al. [21, 23]
		Saari et al. [55]
		Boon et al. [6]
	Banana trunk	Wan Nadhari et al. [39]
	Banana leaves	Nongman et al. [42]
	Rice straw	Kurokochi and Sato [30]
		Jamaludin et al. [24]
	Bagasse	Jamaludin et al. [24]
		Wan Nadhari et al. [38]
	Coconut husk	Greer [18]
	Sunflower stem	Mahieu et al. [35]
Crop	Flax shive	Mahieu et al. [35]
	Hemp shive	Almusawi et al. [2]
	Kenaf bast	Xu et al. [74]
	Kenaf core	Okuda and Sato [45]
	Bamboo	Suhasman et al. [60]
	Moso bamboo	Guan et al. [19]
	Sorghum	Ferrandez-Garcia et al. [15]
	Water hyacinth petiole	Salas-Ruiz et al. [57]

grinding and extrusion processes are required to enhance the properties of binderless composites. Subsequently, [6, 55] have done the steam pretreatment of OPT. In a study done by [55], the OPT chips were exposed at three temperature levels of 100, 110 and 130 °C for 10, 30 and 50 min before they were dried to 6–8% MC and pressed using a pressure of 5 MPa at a temperature of 180 °C for 20 min. They revealed that the steaming process resulted in a positive effect on the properties of the binderless particleboard and the optimal condition was given by specimens undergoing steam pretreatment for 30 min at a temperature of 130 °C. Meanwhile, [6] steamed the OPT particles using autoclave at 160 °C and 35 psi for 20, 40, and 60 min, respectively prior to the air dried of the particles. The finding shows that the binderless OPT particleboard became stronger with the increase of steam pretreatment duration. Nevertheless, the physical and mechanical properties of the steam-pretreated binderless particleboard are not comparable with particleboard made from 10% UF resin.

Banana stem and leaves are categorized as agricultural wastes and can be extracted from the banana tree. Wan [39] examined the properties of steam pre-treatment banana trunk binderless particleboard for 15 min at three levels of temperatures; 111, 121, and 131 °C. Particleboard exposed to steam at 121 °C temperature had significantly greater MOR and modulus of elasticity (MOE) as compared to untreated binderless particleboard. In their study, agricultural waste such as banana stem has the potential to be used as raw material for the manufacture of binderless particleboard, whereas they are comparable with the minimum requirement of Japanese Industrial Standards (JIS) A 5908: 2003. The performance of laminated banana leaf tape and binderless banana stem particleboard also have been investigated by [42]. In their works, they utilize both the leaf and stem of *Musa paradisiaca* var. *awak* and the flexural properties of the board are increased as the number of banana leaf tapes increases.

Few studies on binderless particleboard from rice straw were done by [24, 30]. As reported by [24], the mechanical properties of binderless rice straw particleboard namely MOR, MOE, and IB are slightly lower compared to others with values of 13, 1,028, and 0.3 MPa, respectively. These low values might be due to the high ash, low lignin, and high silica contents of rice straw [17, 24]. Kargbo et al. [26] reported that the ash content of rice straw is up to 17%, thus affecting the performance of binderless rice straw particleboard. As reported by [24], The binderless particleboard made from bagasse shows greater mechanical properties than rice straw, and the poor properties of rice straw can be improved by mixing the rice straw with bagasse particles during the manufacturing process.

Mahieu et al. [35] conducted a study to replace wood in particleboard manufacturing using flax shives (the core or woody part of the flax stalks) and sunflower bark (the external part of the sunflower stem), bonded with 16% UF resin (control) and self-bonding (no binder added). As stated by [35], for the manufacturing of binderless particleboards, a high amount of water was sprayed onto the particles (80% of the initial particle mass) to generate steam during the thermocompression process which contributes to extracting and plasticizing the pecto-lignocellulosic compounds contained in the agroresources [8, 48]. These binderless particleboards were pressed at 190 °C for 45 min (for water evaporation). Findings revealed that binderless flax shives particleboard has interesting bending strength but poor internal bond. Additionally, the denser binderless specimens have better resistance to fire and low resistance to water as compared to specimens with synthetic binder.

Hemp shive which is categorized as waste after extraction of hemp fibres can be used for insulation mats or lightweight concrete. Almusawi et al. [2] thermoformed the grounded hemp shive at four different temperatures of 130, 150, 170 and 190 °C, at a pressure of 20 MPa for 30 min without any synthetic resin. Then, the boards were cold down until reached a temperature of 60 °C prior to mold release. The optimum boards are given by binderless particleboard pressed at a temperature of 170 °C with maximum flexural strength of 27 MPa and MOE of 3122 MPa and meet the minimum requirement of NF EN 312:P6 standard. However, hemp shive binderless particleboard had poor resistance towards water. Apart from hemp, the properties of kenaf (bast and core) binderless particleboard also have been evaluated in prior studies [45, 74]. In a study done by [74], kenaf composites were manufactured using kenaf bast fiber-woven sheets at face and back surfaces, while kenaf core particles are located at the core part. During the manufacturing process, the furnishings are pressed into composite with a sealed steam-injection press and the mats were pressed under the steam treatment with a pressure of 1.0 MPa (183 °C). The finding shows that the binderless kenaf composite has a low internal bond strength (IB) value due to the decrease in core region density. However, the composites with densities of 0.42 and 0.54 g/cm^3 meet the requirements for type 13 and type 18 particleboard, respectively, for both MOR and IB values.

Suhasman et al. [60] investigated three species of Sulawesi bamboo, namely *Gigantochloa ater, Dendrocalamus asper* and *Schizostacyum barcahycladum.* Prior to the hot press, the bamboo particles are oxidized with 15% hydrogen peroxide (based on dry particle weight) and 7.5% of ferrous sulfate (based on the weight of hydrogen peroxide). Then, the mat was hot-pressed at a temperature of 180 °C for 12 min. The highest MOR, MOE and IB values are given by composites made from *Dendrocalamus asper, Schizostacyum barcahycladum* and *Dendrocalamus asper,* respectively. They also indicated that the application of the oxidation method is a success in producing binderless particleboard but unfortunately none of the results have met the JIS A 5908: 2003 requirements. Meanwhile, [19] fermented Moso bamboo using lactic acid bacteria microorganisms prior to the hot pressing. In their finding, the slenderness ratio of flakes decreased at first and then increased with an increase in the fermentation time (in days), and the chemical components of the bamboo green residues degraded with time.

Water hyacinth petiole (WHP) is a rapid growth of invasive aquatic plant species. Salas-Ruiz et al. [57] reported that the process of manufacturing this aquatic plant differs from standard binderless particleboard, where it has no hot-press process. In their study, WHP particles were mixed with water for 300 s, with water to WHP with a ratio of 1.75, then poured into a mould and compressed at 40 Pa for 24 h. The application of WHP binderless particleboard is towards thermal insulation board thus requires low strength of composite. In conclusion, both agricultural waste and crops gained attention from the researchers. Apart from their availability, particleboard is easier to manufacture compared to fibreboard. Overall, the steam treatment process resulted in a positive effect on the properties of the binderless particleboard and improved the overall properties.

4 Binderless Plywood

Binderless plywood has emerged as an eco-friendly alternative to traditional plywood as it was invented without the use of synthetic adhesives. Its reduced environmental impact has garnered significant attention in recent years [10, 27]. The manufacturing process of binderless plywood involves the fusion of wood fibres using a combination of heat and pressure, a method referred to as thermo-mechanical compression. The wood fibres are compressed together at elevated temperatures, thereby softening the lignin present in the wood, which acts as a natural adhesive and binds the fibres together. Importantly, the absence of synthetic adhesives in the manufacturing process ensures that binderless plywood does not emit harmful chemicals or volatile organic compounds (VOCs) into the environment. This not only makes it a safer and healthier option for manufacturers but also for consumers.

According to the Scopus database, the number of published documents on binderless plywood is relatively lower compared to binderless fibreboard and particleboard. Specifically, binderless fibreboard and particleboard have been the focus of 6 and 10 times more published documents over the years, respectively. This discrepancy in research focus may be attributed to the intended applications of these panels. While fibreboard and particleboard are primarily used for non-structural purposes, plywood is often utilized in structural applications. As such, a higher level of bond integrity is required to ensure the overall safety and durability of the product, especially in the face of fluctuating environmental conditions.

Although research on binderless plywood is less extensive compared to binderless fibreboard and particleboard, several studies have reported that delamination under moist or wet conditions is a main issue for binderless plywood [29, 52]. As a hygroscopic material, the veneers in a plywood panel can absorb and desorb moisture due to the abundance of hydroxyl groups in their chemical composition. Therefore, research on binderless plywood has focused on enhancing the bondability between veneer layers through veneer pretreatment [27, 29, 52, 54], post-production treatments to further improve properties [52], and manipulating processing parameters to optimize and activate chemical constituents as self-binders [10, 27, 56].

A method was developed to produce boards from wood particles and fibres without synthetic binders using hydrogen peroxide and ferrous sulfate as catalysts [27, 54]. The resulting boards exhibited higher resistance to water swelling compared to untreated boards. The bondability of the treated wood was found to be attributed to the formation of extractives by oxidative treatment, with long-lived bonding groups being crucial for bond formation. Leaching the treated wood with water removed most of the bondability, but re-adding the leachate followed by drying recreated much of the bondability.

To produce binderless plywood, [27] sprayed an aqueous solution of ferrous sulfate (0.1% odw) followed by 0.5% hydrogen peroxide onto Scots pine veneer prior to conditioning at 20 °C and 60% relative humidity and hot pressing. Panels with oxidative-pretreatment veneers exhibited increased stability, with very few specimens showing slight delamination after a 24-h water uptake test. A similar trend was observed for the bending properties of the treated panels in the same study. Alternative methods for pretreatment of veneers without the use of chemicals include wetting veneers before hot pressing and heat treatment after hot pressing, which have been reported to improve bond integrity and decrease delamination when samples were soaked in water. Moisture was found to enhance the formability of lignocellulose materials by improving the mobility of the lignin-hemicellulose matrix and increasing the contact area. Ruponen et al. [52] conducted a post-treatment study on binderless plywood, in which conditioned plywood panels were further thermally modified at a target temperature of 200 °C and atmospheric pressure for 4 h using superheated steam. The post-heat-treated specimens exhibited lower moisture content throughout the moisture absorption test, with delamination almost eliminated compared to untreated specimens. This was likely due to lowered water absorption capacity, relaxation of stored strain energy, and the formation of crosslinks within the lignin-hemicellulose matrix.

Several studies have demonstrated the efficacy of wetting veneers prior to hot pressing, which can enhance the bondability of resulting panels. Ruponen et al. [52] conducted a study using veneers conditioned in two different environments, namely, humid air (20 °C, 60% relative humidity) and water immersion at 20 °C, to fabricate 8-ply binderless birch plywood. The study reported that plywood made from initially wet veneers exhibited greater bond stability after a water uptake test. This phenomenon may be attributed to the increased crosslinks as the lignin-hemicellulose matrix becomes more mobile in a wet condition. However, another study utilizing Norway spruce veneers reported that the veneers had blown during hot pressing due to high veneer moisture [53]. This occurrence may be due to differences in veneer permeability, as spruce has higher permeability and thicker veneer compared to birch veneer.

Kurokochi et al. [29] further explored the use of wet veneers in fabricating 3-ply plywood from sugi (Japanese cedar). The veneers were soaked in tap water at 20 °C for 12 h, while the other veneers were wet under running water for a minute. Both veneer conditions had 50–60 and 12% moisture content, respectively, before the scratching pretreatment and hot-pressing steps. They reported that both pretreatment veneer conditions exhibited high wet shear strength, which was comparable to the dry shear strength of the resulting panels. With regard to scratching pretreatment, scratching with a wire brush required less effort compared to the sandpaper method. Scratching pretreatment alone was not sufficient, but when both wetting and scratching pretreatments were applied, plywood with high self-bonding properties could be achieved. In summary, wetting veneers prior to hot pressing can improve bondability and enhance the crosslinks between the lignin-hemicellulose matrix. However, differences in veneer permeability should be considered to avoid veneer blow during hot pressing. Additionally, the combination of wetting and scratching pretreatments can result in high self-bonding property of plywood.

The production of binderless plywood involves various processing variables, including pressing temperature, pressure, and time, which influence the chemical and physical processes that contribute to self-binding between veneers. Achieving the target temperature in all bond lines is essential to obtain the desired properties of the panel. Different studies have investigated the optimal pressing conditions for various wood species and veneer thicknesses. Saari et al. [56] used response surface methodology to optimize the processing variables for producing 5-layer binderless veneer panels made from oil palm trunks. The study revealed that the pressing temperature and time significantly affect the properties of compressed panels, with an optimum condition of 186.6 °C for 30 min resulting in high bending strength and shear block strength with a low percentage of delamination properties.

Similarly, [10] investigated 15 combinations of pressing parameters, including temperature (200–250 °C), pressure 4–6 MPa and pressing time (240–360 s) for 5-ply binderless plywood from beech veneers. The study showed that severe pressing conditions, particularly at 250 °C, resulted in higher density, higher shear and bending strength, and lower water absorption after 48 h of testing. The panels also did not delaminate at all after the test. Additionally, the pressing cycle depends on the pretreatment used, with [29] reporting that a 4-step pressing cycle is more suitable for

24-h-soaked veneer, while a simple 1-step pressing is appropriate for 1-min veneer wetting treatment, as it can release vapor gradually and avoid explosive delamination during hot pressing.

Density is an essential indicator of the quality of a composite, and it correlates with the panels' performances. Saari et al. [55] fabricated plywood of Sengon wood with three different thicknesses (7, 10, and 15 mm) while maintaining the other processing variables. The thinnest panels resulted in better overall performance due to the good quality created during the compression process. The compression process reduces the total volume of the panel, increasing the density and strength of the panel, as supported by Saad. Therefore, optimizing the pressing conditions while considering the wood species, veneer thickness, and pretreatment is crucial to producing high-quality binderless plywood with desirable properties.

5 Issue and Challenges

Binderless composites have earned interest in the composite sector mostly because of environmental concerns. Apart from the renewable and green materials, the challenges in manufacturing and application of binderless composite are still disputed. Even though there are many studies that have been done on binderless composites, there is still a challenge to scale up the production and commercialize the binderless composites [64, 77]. The most challenging aspect of using lignocellulosic materials, namely fibres, particles and veneers are the self-adhesive mechanism. For instance, fibre must deform to create fibre-to-fibre contact and a strong adhesive network. Increased contact area with fibre polymers as a result of deformation, which operates under the least amount of pressure, is referred to as the rubbery state. The mechanical qualities of phenolic-based fibreboards are often not achieved by binderless or auto-adhesion fibreboards. Thus, binderless fibreboard's mechanical characteristics should be improved at a temperature greater than the glass transition temperature (Tg) of all fibre polymers by increasing the pressing temperature to soften the polymer and increase the contact area. In order to get better self-bonding, binderless fibreboards and particleboards should be manufactured using an integrated process that includes enzyme-mediated pre-treatment techniques including chemical oxidation and steam injection/explosion to overcome some issues including dimensional stability and the impact of internal bonding caused by moisture content.

Apart from that, the cost and the performance of binderless composites are not competitive with commercial resin-based composites [77]. Compared to conventional composites, binderless composites generally exhibit lower mechanical strength and stiffness. However, there are two alternative ways to improve the mechanical properties of binderless composites namely increasing the pressing temperature and enzyme or chemical pre-treatments. Many researchers worked on self-bonded materials to develop binderless composites made up of wood, non-wood, agricultural waste, industrial crops and grass fibre. The chemical makeup, biological and physical characteristics of lignocellulosic material initially varied from one another. To provide high-quality binderless composite each material should have its specific processing parameters defined during board preparation. The material properties, pressing temperature, and time variations throughout the production process have a substantial impact on the mechanical and physical properties.

Similar to fibreboard and particleboard, there are still some challenges and issues with binderless plywood production. One challenge is the high energy consumption required during the manufacturing process, which can make it less economically feasible [52]. Additionally, the strength and durability of the plywood may not be as consistent as traditional plywood, which can limit its applications in certain construction projects. The technology for producing binderless plywood is still relatively new and may require further development and optimization to improve its overall performance and consistency. In summary, the challenges and issues of these binderless composites still need to be explored and further studies on the pressing temperature and pre-treatment should be focused in order to increase their mechanical properties. It could be an alternative potential and accepted by the industry to partially replace conventional composites. Indeed, binderless composites have the potential to replace the conventional composites that use adhesive mostly for non-structural components and furniture. As stated by [9], binderless composites have the potential to be applied to high-strength outdoor household materials and structural buildings, which corresponds with the notions of carbon storage, eco-friendliness and degradation.

Even though there is much research that has been carried out on binderless composites, the studies are still limited to the laboratory scale. As reported by [41], poor mechanical properties and lower stability of those binderless boards have limited their application in the industries. Due to their lightweight properties and less formaldehyde emission, binderless composites are suitable materials for indoor applications. Apart from wall partition walls, it can be used for thermal insulation boards [47, 57].

6 Conclusions

Studies on binderless composites have been done on various types of lignocellulosic materials mostly for fibreboard, particleboard and plywood. Binderless composites have been recognized as an environmentally friendly alternative to traditional composites as they are manufactured without the use of synthetic adhesives. Five subtopics have been discussed namely introduction, binderless fibreboard, binderless particleboard, binderless plywood and issues and challenges. In the first topic, explanations of binderless composites, especially on the manufacturing process are broadly discussed. In the next three subtopics, the materials, parameters and properties are described, while the last subtopic elaborates on the issue, challenges, application and potential research of binderless composites. In terms of materials, binderless fibreboards can be made from various categories of plants namely wood, agricultural waste, crops, and grass. Binderless particleboards are limited to agricultural waste and crops, while plywood is only suitable for using wood species and oil palm trunk. A diagram of the manufacturing processes of binderless composites is developed, and the differentiations of conventional composite and binderless composite are discussed. In the manufacturing of binderless composites, the pressing times and pressing temperatures are generally higher compared to conventional composites. Mostly, increasing pressing time and temperature reduces the water absorption and improves the dimension stability of binderless composite. In conclusion, binderless composites can be sustainable alternative materials to conventional composites in various applications with further exploration in terms of pre-treatment and the parameters. With ongoing research and development efforts, binderless biocomposites offer several advantages, including reduced environmental impact and improved biodegradability of the product.

Acknowledgements The authors would like to acknowledge Universiti Tun Hussein Onn Malaysia and Universiti Putra Malaysia for their support and encouragement.

References

1. Abbott, A.P., Conde, J.P., Davis, S.J., Wise, W.R.: Starch as a replacement for urea-formaldehyde in medium density fibreboard. Green Chem. **14**(11), 3067–3070 (2012)
2. Almusawi, A., Lachat, R., Atcholi, K.E., Gomes, S.: Proposal of manufacturing and characterization test of binderless hemp shive composite. Int. Biodeterior. Biodegrad. **115**, 302–307 (2016)
3. Araújo Junior, C.P., Coaquira, C.A.C., Mattos, A.L.A., de Souza Filho, M.D.S.M., Feitosa, J.P.D.A., Morais, J.P.S.D., de Freitas Rosa, M.: Binderless fibreboards made from unripe coconut husks. Waste Biomass Valoriz. **9**, 2245–2254 (2018)
4. Bahari, S.A., Jamaludin, M.A., Ahmad, M., Subardi, S.A., Saad, M.J., Nordin, K.: Characterization on bending strength of binderless bamboo fibreboard" (BBF). In: Proceeding of 9th Pacific Rim Bio-based Composites Symposium, CSIRO and SCION, New Zealand, pp. 238–246 (2008)
5. Baskaran, M., Hashim, R., Said, N., Raffi, S.M., Balakrishnan, K., Sudesh, K., Sulaiman, O., Arai, T., Kosugi, A., Mori, Y., Sugimoto, T., Sato, M.: Properties of binderless particleboard from oil palm trunk with addition of polyhydroxyalkanoates. Compos. B Eng. **43**(3), 1109–1116 (2012)
6. Boon, J.G., Hashim, R., Danish, M., Nadhari, W.N.A.W.: Physical and mechanical properties of binderless particleboard made from steam-pretreated oil palm trunk particles. J. Compos. Sci. **3**(2), 46 (2019)
7. Boran, S., Usta, M., Gümüşkaya, E.: Decreasing formaldehyde emission from medium density fibreboard panels produced by adding different amine compounds to urea formaldehyde resin. Int. J. Adhes. Adhes. **31**(7), 674–678 (2011)
8. Bouajila, J., Limare, A., Joly, C., Dole, P.: Lignin plasticization to improve binderless fibreboard mechanical properties. Polym. Eng. Sci. **45**(6), 809–816 (2005)
9. Chen, Y., Zhang, Y., Cao, Y., Guo, D., Li, G., Qing, Y., Wu, Y.: High-performance self-bonding bio-composites from wood fibers. Ind. Crops Prod. **209**, 117944 (2024)
10. Cristescu, C., Sandberg, D., Ekevad, M., Karlsson, O.: Influence of pressing parameters on mechanical and physical properties of self-bonded laminated beech boards. Wood Mat. Sci. Eng. **10**(2), 205–214 (2015)

11. Van Dam, J.E., van den Oever, M.J., Teunissen, W., Keijsers, E.R., Peralta, A.G.: Process for production of high density/high performance binderless boards from whole coconut husk: part 1: lignin as intrinsic thermosetting binder resin. Ind. Crops Prod. **19**(3), 207–216 (2004)
12. Delgado-Aguilar, M., González, I., Pèlach, M.A., De La Fuente, E., Negro, C., Mutjé, P.: Improvement of deinked old newspaper/old magazine pulp suspensions by means of nanofibrillated cellulose addition. Cellulose **22**, 789–802 (2015)
13. Domínguez-Robles, J., Tarres, Q., Alcala, M., El Mansouri, N.E., Rodríguez, A., Mutje, P., Delgado-Aguilar, M.: Development of high-performance binderless fibreboards from wheat straw residue. Constr. Build. Mater. **232**, 117247 (2020)
14. Espinosa, E., Tarrés, Q., Theng, D., Delgado-Aguilar, M., Rodríguez, A., Mutjé, P.: Effect of enzymatic treatment (endo-glucanases) of fibre and mechanical lignocellulose nanofibres addition on physical and mechanical properties of binderless high-density fibreboards made from wheat straw. J. Build. Eng. **44**, 103392 (2021)
15. Ferrandez-Garcia, A., Ferrandez-Garcia, M.T., Garcia-Ortuño, T., Ferrandez-Villena, M.: Influence of the density in binderless particleboards made from sorghum. Agronomy **12**(6), 1387 (2022)
16. Gao, Z., Wang, X. M., Wan, H., Brunette, G.: Binderless panels made with black spruce bark. BioResources **6**(4) (2011)
17. Ghaffar, S.H., Fan, M.: Lignin in straw and its applications as an adhesive. Int. J. Adhes. Adhes. **48**, 92–101 (2014)
18. Greer, S.: Converting coconut husks into binderless particle board. Doctoral dissertation (2009)
19. Guan, M., Fu, R., Yong, C., Li, Y., Xu, X.: Properties of binderless bamboo particleboards derived from biologically fermented bamboo green residues. Waste Manag. **151**, 195–204 (2022)
20. Halvarsson, S., Edlund, H., Norgren, M.: Properties of medium-density fibreboard (MDF) based on wheat straw and melamine modified urea formaldehyde (UMF) resin. Ind. Crops Prod. **28**(1), 37–46 (2008)
21. Hashim, R., Nadhari, W.N.A.W., Sulaiman, O., Kawamura, F., Hiziroglu, S., Sato, M., Tanaka, R.: Characterization of raw materials and manufactured binderless particleboard from oil palm biomass. Mater. Des. **32**(1), 246–254 (2011)
22. Hashim, R., Saari, N., Sulaiman, O., Sugimoto, T., Hiziroglu, S., Sato, M., Tanaka, R.: Effect of particle geometry on the properties of binderless particleboard manufactured from oil palm trunk. Mater. Des. **31**(9), 4251–4257 (2010)
23. Hashim, R., Said, N., Lamaming, J., Baskaran, M., Sulaiman, O., Sato, M., Sugimoto, T.: Influence of press temperature on the properties of binderless particleboard made from oil palm trunk. Mater. Des. **32**(5), 2520–2525 (2011)
24. Jamaludin, M.A., Bahari, S.A., Zakaria, M.N., Saipolbahri, N.S.: Influence of rice straw, bagasse, and their combination on the properties of binderless particleboard. J. Korean Wood Sci. Technol. **48**(1), 22–31 (2020)
25. Juliana, A.H., Lee, S.H., Paridah, M.T., Ashaari, Z., Lum, W.C.: Development and characterization of wood and non-wood particle based green composites. In: Green Biocomposites: Manufacturing and Properties, pp. 181–198 (2017)
26. Kargbo, F.R., Xing, J., Zhang, Y.: Property analysis and pretreatment of rice straw for energy use in grain drying: a review. Agric. Biol. JN Am **1**(3), 195–200 (2010)
27. Karlsson, O., Cristescu, C., Sandberg, D.: Autoadhesion of laminated boards from scots pine veneers: effect of oxidative pretreatment 11 (2015)
28. Kharazipour, A., Bergmann, K., Nonninger, K., Ludemann, H.D.: Properties of fibre boards obtained by activation of the middle lamella lignin of wood fibres with peroxidase and H2O2 before conventional pressing. J. Adhes. Technol. **12**(19), 1045–1053 (1998)
29. Kurokochi, Y., Hasegawa, W., Sato, M.: The effects of wetting and scratching pretreatment of veneers on shear strength of binderless plywood made from sugi (Japanese cedar, Cryptomeria japonica). J. Wood Sci. **65**(1), 15 (2019)
30. Kurokochi, Y., Sato, M.: Properties of binderless board made from rice straw: the morphological effect of particles. Ind. Crops Prod. **69**, 55–59 (2015)

31. Laemsak, N., Okuma, M.: Development of boards made from oil palm frond II: properties of binderless boards from steam-exploded fibres of oil palm frond. J. Wood Sci. **46**, 322–326 (2000)
32. Lee, T.C., Pu'ad, N.M., Selimin, M.A., Manap, N., Abdullah, H.Z., Idris, M.I.: An overview on development of environmental friendly medium density fibreboard. Mater. Today: Proc. **29**, 52–57 (2020)
33. Luo, H., Yue, L., Wang, N., Zhang, H., Lu, X.: Manufacture of binderless fibreboard made from bamboo processing residues by steam explosion pretreatment. Wood Res. **59**(5), 861–870 (2014)
34. Luthfi, N., Wang, X., Kito, K.: Effect of drying temperature on the physical properties of binderless fibreboard from bagasse: study of water absorption. Sci. & Technol. Asia 30–38 (2021)
35. Mahieu, A., Alix, S., Leblanc, N.: Properties of particleboards made of agricultural by-products with a classical binder or self-bound. Ind. Crops Prod. **130**, 371–379 (2019)
36. Mancera, C., Ferrando, F., Salvado, J.: Cynara cardunculus as raw material for the production of binderless fiberboards: optimization of pretreatment and pressing conditions. J. Wood Chem. Technol. **28**(3), 207–226 (2008)
37. Mejía, E.H., Quintana, G.C., Ogunsile, B.O.: Development of binderless fibreboards from steam exploded and oxidized oil palm wastes. BioResources **9**(2), 2922–2936 (2014)
38. Nadhari, W.N.A.W., Abd Karim, N., Boon, J.G., Salleh, K.M., Mustapha, A., Hashim, R., Azni, M.E.: Sugarcane (Saccharum officinarium L.) bagasse binderless particleboard: Effect of hot pressing time study. Mater. Today: Proc. **31**, 313–317 (2020)
39. Nadhari, W.N.A.W., Danish, M., Nasir, M.S.R.M., Geng, B.J.: Mechanical properties and dimensional stability of particleboard fabricated from steam pre-treated banana trunk waste particles. J. Build. Eng. **26**, 100848 (2019)
40. Nasir, M., Gupta, A., Beg, M.D.H., Chua, G.K., Jawaid, M., Kumar, A., Khan, T.A.: Fabricating eco-friendly binderless fibreboard from laccase-treated rubber wood fibre. BioResources **8**(3), 3599–3608 (2013)
41. Nasir, M., Khali, D.P., Jawaid, M., Tahir, P.M., Siakeng, R., Asim, M., Khan, T.A.: Recent development in binderless fibre-board fabrication from agricultural residues: a review. Constr. Build. Mater. **211**, 502–516 (2019)
42. Nongman, A.F., Baharin, A., Bakar, A.A.: The effect of banana leaves lamination on the mechanical properties of particle board panel. Procedia Chem. **19**, 943–948 (2016)
43. Okuda, N., Hori, K., Sato, M.: Chemical changes of kenaf core binderless boards during hot pressing (I): influence of the pressing temperature condition. J. Wood Sci. **52**, 244–248 (2006)
44. Okuda, N., Hori, K., Sato, M.: Chemical changes of kenaf core binderless boards during hot pressing (II): effects on the binderless board properties. J. Wood Sci. **52**, 249–254 (2006)
45. Okuda, N., Sato, M.: Manufacture and mechanical properties of binderless boards from kenaf core. J. Wood Sci. **50**, 53–61 (2004)
46. Okuda, N., Sato, M.: Water resistance properties of kenaf core binderless boards. J. Wood Sci. **52**(5), 422–428 (2006)
47. Panyakaew, S., Fotios, S.: New thermal insulation boards made from coconut husk and bagasse. Energy Build. **43**(7), 1732–1739 (2011)
48. Pintiaux, T., Viet, D., Vandenbossche, V., Rigal, L., Rouilly, A.: Binderless materials obtained by thermo-compressive processing of lignocellulosic fibers: a comprehensive review. BioResources **10**(1) (2015)
49. Puspaningrum, T., Haris, Y.H., Sailah, I., Yani, M., Indrasti, N.S.: Physical and mechanical properties of binderless medium density fibreboard (MDF) from coconut fibre. IOP Conf. Ser.: Earth Environ. Sci. **472**(1):012011. IOP Publishing (2020)
50. Quintana, G., Velasquez, J., Betancourt, S., Ganan, P.: Binderless fibreboard from steam exploded banana bunch. Ind. Crops Prod. **29**(1), 60–66 (2009)
51. Ramesh, M., RajeshKumar, L., Bhuvaneshwari, V.: Bamboo fibre reinforced composites. In: Bamboo Fibre Composites: Processing, Properties and Applications, pp. 1–13 (2021)

52. Ruponen, J., Rautkari, L., Belt, T., Hughes, M.: Factors influencing properties of parallel laminated binderless bonded plywood manufactured from rotary cut birch (Betula pendula L.). Int. Wood Prod. J. **5**(1), 11–17 (2014)
53. Ruponen, J., Rautkari, L., Hughes, M.: Enhancing the Properties of Plywood Produced by Auto-Adhesion, pp. 331–338 (2010)
54. Saad, S., Yunianti, A.D., Suhasman, S.: Effect of layer structure on physical and mechanical properties of binderless composite plywood. IOP Conf. Ser.: Mater. Sci. Eng. **593**(1), 012012 (2019)
55. Saari, N., Hashim, R., Sulaiman, O., Hiziroglu, S., Sato, M., Sugimoto, T.: Properties of steam treated binderless particleboard made from oil palm trunks. Compos. B Eng. **56**, 344–349 (2014)
56. Saari, N., Lamaming, J., Hashim, R., Sulaiman, O., Sato, M., Arai, T., Kosugi, A., Wan Nadhari, W.N.A.: Optimization of binderless compressed veneer panel manufacturing process from oil palm trunk using response surface methodology. J. Clean. Prod. **265**, 121757 (2020)
57. Salas-Ruiz, A., del Mar Barbero-Barrera, M., Ruiz-Téllez, T.: Microstructural and thermo-physical characterization of a Water Hyacinth petiole for thermal insulation particle board manufacture. Materials **12**(4), 560 (2019)
58. Shi, J., Zhong, T., Xu, X., Wu, J., Zou, Y., Fei, B., Xia, C., Chen, H.: Eco-friendly and special-shaped bamboo binderless fibreboards fabricated by self-bonding technology: effect of bamboo fibres with different sizes. Ind. Crops Prod. **194**, 116300 (2023)
59. Suhaily, S.S., Jawaid, M., Abdul Khalil, H.P.S., Mohamed, A.R., Ibrahim, F.: A review of oil palm biocomposites for furniture design and applications: potential and challenges. BioResources **7**(3), 4400–4423 (2012)
60. Suhasman, S., Yunianti, D., Saad, S., Baharuddin, B.: Characteristics of binderless particleboard made of three species of sulawesi bamboos. Wood Res. J. **4**(2), 68–71 (2013)
61. Tajuddin, M., Ahmad, Z., Ismail, H.: A review of natural fibres and processing operations for the production of binderless boards. BioResources **11**(2), 5600–5617 (2016)
62. Theng, D., Arbat, G., Delgado-Aguilar, M., Vilaseca, F., Ngo, B., Mutjé, P.: All-lignocellulosic fibreboard from corn biomass and cellulose nanofibres. Ind. Crops Prod. **76**, 166–173 (2015)
63. Velasquez, J.A., Ferrando, F., Farriol, X., Salvado, J.: Binderless fibreboard from steam exploded Miscanthus sinensis. Wood Sci. Technol. **37**(3–4), 269–278 (2003)
64. Vitrone, F., Ramos, D., Ferrando, F., Salvadó, J.: Binderless fibreboards for sustainable construction materials, production methods and applications. J. Build. Eng. **44**, 102625 (2021)
65. Vitrone, F., Ramos, D., Ferrando, F., Salvadó, J.: The influence of thickness and final heat treatment on mechanical properties and dimensional stability of binderless fibreboards from steam exploded Arundo donax L. Mater. Today: Proc. **58**, 999–1004 (2022)
66. Wang, J., Liu, J., Ni, L., Li, J.: Thermopressed binderless fibreboards from wheat straw by adding black liquor. Wood Res **63**, 1021–1032 (2018)
67. Wang, J., Wang, B., Liu, J., Ni, L., Li, J.: Effect of hot-pressing temperature on characteristics of straw-based binderless fibreboards with pulping effluent. Materials **12**(6), 922 (2019)
68. Widyorini, R., Xu, J., Umemura, K., Kawai, S.: Manufacture and properties of binderless particleboard from bagasse I: effects of raw material type, storage methods, and manufacturing process. J. Wood Sci. **51**, 648–654 (2005)
69. Widyorini, R., Xu, J., Watanabe, T., Kawai, S.: Chemical changes in steam-pressed kenaf core binderless particleboard. J. Wood Sci. **51**, 26–32 (2005)
70. Wu, J., Wang, Y., Zhang, Z., Lu, X., Wang, C., Zhong, Z., Qian, D., Zhang, T.: Optimization of pretreating poplar wood shavings with cellulase to produce binderless fibreboard using response surface methodology. BioResources **15**(3), 6431–6441 (2020)
71. Xie, L., Liu, J., Du, A.: Effect of hot-pressing factors on binderless fibreboard properties. In: Proceedings of 2012 International Conference on Biobase Material Science and Engineering. IEEE, October 2012, pp. 8–11 (2012)
72. Xu, J., Han, G., Wong, E.D., Kawai, S.: Development of binderless particleboard from kenaf core using steam-injection pressing. J. Wood Sci. **49**, 327–332 (2003)

73. Xu, J., Sugawara, R., Widyorini, R., Han, G., Kawai, S.: Manufacture and properties of low-density binderless particleboard from kenaf core. J. Wood Sci. **50**, 62–67 (2004)
74. Xu, J., Widyorini, R., Kawai, S.: Properties of kenaf core binderless particleboard reinforced with kenaf bast fibre-woven sheets. J. Wood Sci. **51**(4), 415–420 (2005)
75. Xu, J., Widyorini, R., Yamauchi, H., Kawai, S.: Development of binderless fibreboard from kenaf core. J. Wood Sci. **52**, 236–243 (2006)
76. Yang, Z., Song, W., Cao, Y., Wang, C., Hu, X., Yang, Y., Zhang, S.: The effect of laccase pretreatment conditions on the mechanical properties of binderless fibreboards with wheat straw. BioResources **12**(2), 3707–3719 (2017)
77. Zhang, D., Zhang, A., Xue, L.: A review of preparation of binderless fibreboards and its self-bonding mechanism. Wood Sci. Technol. **49**, 661–679 (2015)
78. Zhou, X.Y., Zheng, F., Li, H.G., Lu, C.L.: An environment-friendly thermal insulation material from cotton stalk fibers. Energy Build. **42**(7), 1070–1074 (2010)

Lignin-Based Adhesive Bonded Veneer Composites: A Sustainable Approach

Shibu Comath, Arya Chirakkal, and Ashline Shaji

Abstract Over the past two decades, the global wood panel industry has witnessed a substantial increase in production and consumption, driven largely by the construction sector. However, the dominance of synthetic adhesives like phenol-formaldehyde, urea-formaldehyde, and melamine-formaldehyde has raised environmental concerns due to their toxic components derived from fossil resources. There has been a growing interest in bio-based adhesives derived from renewable materials like lignin to address these issues and promote sustainability within the wood panel industry. This chapter explores the potential of lignin-based adhesives, focusing on their development, challenges, and environmental impact, particularly in the context of veneer-based composites.

Keywords Lignin · Formaldehyde-free · Bio-adhesive · Formaldehyde emission · Plywood

1 Introduction

Over the past two decades, the global production and consumption of wood-based materials have grown remarkably, with wood-based materials gaining increasing significance in the construction industry. Synthetic adhesives currently dominate the global wood adhesive market, accounting for an overwhelming 95% share. Asian countries are the largest synthetic adhesive consumers, representing 32% of the global market [34]. Historically, humans have used bio-adhesives, such as plant and collagen-based glues, by ancient Egyptians. However, synthetic adhesives gradually replaced their bio-based counterparts from the 1930s onward due to economic

S. Comath (✉) · A. Shaji
Department of Forest Product and Utilization, College of Forestry, Kerala Agricultural University, Thrissur, Kerala, India
e-mail: shibu-2021-27-011@student.kau.in

A. Chirakkal
Department of Silviculture and Agroforestry, College of Forestry, Kerala Agricultural University, Thrissur, Kerala, India

S. S. Osman Al Edrus et al. (eds.), *Green Lignocellulosic-Based Panels*, Engineering Materials, https://doi.org/10.1007/978-981-96-3404-0_4

considerations, water resistance, and ease of use. Recent years have witnessed a resurgence of interest in bio-based adhesives, driven by concerns over formaldehyde emissions and the growing demand for sustainable and renewable adhesive materials [26]. Researchers have delved into the development of environmentally friendly adhesives sourced from renewable materials such as protein, lignin, tannin, and starch, seeking alternatives to synthetic adhesives.

Among these renewable materials, lignin has a much higher priority than others because lignin is found in the cell walls of plants. It is the second most abundant biopolymer in nature after cellulose. Lignin has several properties that make it a good candidate for use in adhesives, including high thermal stability, hydrophobicity, and resistance to pathogens.

However, the performance of lignin-based adhesives varies due to factors like molecular weight heterogeneity, complex structure, and limited reactivity [3]. Lignin exhibits reduced reactivity due to fewer active sites, functional groups, and steric hindrances, necessitating modification and the addition of crosslinkers to enhance their bonding capabilities in wood adhesives. Therefore, several research studies formulated different kinds of lignin-based wood adhesives in recent years. The primary objective of this chapter is to provide an overview of the history of bio-adhesive, formaldehyde legislation, and veneer composite development bonded with lignin-based adhesives.

2 History of Bio-based Adhesives

The history of bio-based adhesives extends back over 40,000 years, with evidence of their use dating to the Middle Palaeolithic era when Neanderthals utilized birch bark tar for toolmaking, as confirmed through gas chromatography-mass spectrometry (GC-MS) analysis [33]. As time progressed, adhesive techniques and resources underwent significant evolution, leading to the widespread incorporation of adhesives into various technical and artistic processes. Notably, in ancient Egypt, hides were employed to create animal glue, which was first used over 5,130 years ago in the construction of wooden furniture (chair of Hatnefer) [8].

The historical utilization of adhesives was not limited to Egypt, as both the Greeks and Romans harnessed natural materials, including egg white, blood, bones, milk, cheese, vegetables, and grains, for veneer and inlay work in woodworking. In their unprocessed forms, these early bio-adhesives represented the initial generation of low-strength adhesives primarily suited for interior furniture applications.

However, the dynamics of adhesive usage transformed during significant historical periods, such as the World Wars and the Industrial Revolution, and these pivotal moments spurred the development of readily available and cost-effective synthetic adhesives, gradually supplanting their bio-based predecessors in woodworking applications. The emergence of phenol formaldehyde in the 1930s marked a critical milestone as the first synthetic wood adhesive, revolutionizing the production of exterior plywood.

World War II played a pivotal role in advancing the field of synthetic adhesives, with the introduction of various formulations, including polyvinyl acetate, polyurethane epoxies, and aliphatic resins, all contributing to the growth and diversification of adhesive technology [3]. During World War II, the Germans used a clever trick. They made fake wooden planes and military equipment to trick the British. The idea was to make the British waste their time and resources attacking these fake things so they would not have as much left to attack the real German targets. When the British caught on to this trick, they came up with a funny response. They dropped wooden dummy bombs on the fake German stuff, and these dummy bombs had the words "Wood for Wood" written on them. This was their way of making fun of the German tactic and letting the German soldiers know they were not fooled [5].

Starting from the 1970s, scientists worldwide became more interested in making adhesives for wood from materials that can be renewed. This interest grew because of an oil crisis, but it slowed down when oil became cheaper again. However, in the twenty-first century, the interest came back, and there are a few reasons why. One big reason is that people care much more about protecting the environment now. Also, governments have made new rules to protect the environment, and more people are thinking about how to be kinder to our planet [32].

3 Formaldehyde Legislation

Formaldehyde is a clear, flammable gas with a strong smell. It applies to various products, including wood composites, pressed wood items, construction materials, and household goods. Formaldehyde is a well-known carcinogen to humans and can lead to several health issues, such as respiratory irritation, allergies, and cancer. Formaldehyde was initially synthesized in 1885 and identified as one of the indoor air pollutants in 1980. In 2004, the International Agency for Research on Cancer (IARC) classified formaldehyde as a human carcinogen. By 2010, the World Health Organization (WHO) established a guideline value for formaldehyde at 0.1 mg/m^3 (0.08 ppm). The United States enacted its first formaldehyde legislation in 1978 through the Consumer Product Safety Commission (CPSC), which imposed limits on formaldehyde emissions from pressed wood products. In 1980, the Environmental Protection Agency (EPA) established the National Ambient Air Quality Standard (NAAQS) for formaldehyde, which governs the allowable presence of formaldehyde in outdoor air [24, 63]. Recent years have witnessed the implementation of formaldehyde regulations in other nations. Japan has introduced minimum exposure levels for formaldehyde in workplaces. Numerous countries have formulated guidelines or protocols for formaldehyde use, including ASTM, JIS, CARB, and ANIS [60]. However, there are currently no mandatory regulations in India governing formaldehyde emissions [54].

4 Lignin: The World's Newest Obsession

Wood composites are produced by bonding together wood veneers or particles. Historically, formaldehyde-based adhesives were the conventional choice for this purpose. However, formaldehyde is a well-known carcinogen, leading to an increasing demand for bio-based adhesives that prioritize safety and sustainability. Bio-adhesives are crafted from renewable resources like plant proteins, starch, and lignin. They typically possess non-toxic and biodegradable characteristics, making them an eco-friendly alternative to formaldehyde-based adhesives.

In recent years, interest in bio-adhesives for wood composites has surged. Several factors contribute to this trend, including heightened awareness of the health hazards of formaldehyde, a growing market for sustainable products, and advancements in bio-adhesive technologies. Biopolymers derived from renewable sources, such as plants, animals, or microorganisms, are integral to this innovation. They offer a sustainable substitute for petroleum-based polymers and boast numerous advantages, including biodegradability, renewability, and low toxicity [58]. While bio-adhesives represent a relatively novel technology, they hold immense potential to transform the wood composites industry.

Bio-based adhesives are made from biopolymers such as proteins, tannins, carbohydrates, and lignin. Among these polymers, lignin has a much higher priority than others because lignin is found in the cell walls of plants. It is the second most abundant biopolymer in nature after cellulose. Lignin has several properties that make it a good candidate for use in adhesives, including high thermal stability, hydrophobicity, and resistance to pathogens. There are two main types of lignin-based adhesives (Fig. 1): Lignin-Formaldehyde (LF) adhesives and Lignin-Formaldehyde-Free (LFF) adhesives. Lignin-based adhesives offer several advantages over conventional petroleum-based adhesives, including biodegradability, renewability, and low toxicity [24].

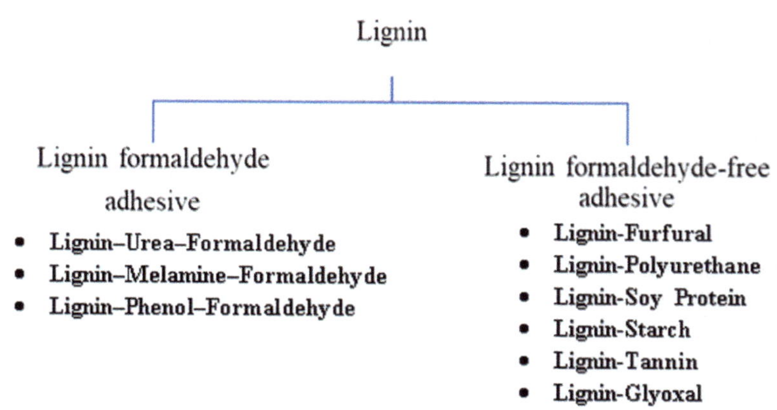

Fig. 1 Classification of lignin-derived wood adhesives

Lignin background and chemistry

Lignin is composed of phenylpropane units. The structure of lignin is not well-defined, and it cannot be isolated in its native, unaltered state. Lignin comprises three fundamental monolignols: sinapyl, coniferyl, and p-coumaryl alcohols. However, the proportion of these monomers varies across diverse plant species. The types of lignin can be divided into softwood lignin, which mainly consists of coniferyl units; hardwood lignin, which consists of coniferyl-sinapyl units; and grass lignin, which consists of coniferyl-sinapyl units. The reactivity of lignin largely depends on its linkages and functional groups. Functional groups, including alcohol hydroxyl, phenolic hydroxyl, carbonyl, carboxyl, and methoxyl groups, are important structural characteristics of lignin. These functional groups directly determine its optical properties, dispersion characteristics, and chemical reactivity. The chemical reactivity of lignin is limited because the linkages connecting the phenylpropane units include C–O and C–C bonds. Lignin contains a very low number of free positions in the lignin C_9 units. Lower reactivity is also attributed to the less reactive positions of C_6 and C_2 in phenol [24, 32]. Lignin can be classified into two categories: native lignin and technical lignin. Native lignin is the lignin found in plants in its natural form, whereas technical lignin is lignin extracted from plants for further processing. There are several ways to extract technical lignin from biomass, such as sulfur pulping, sulfur-free pulping, and biorefinery processes. There are only three practices in the industry: sulfite pulping, kraft pulping, and soda pulping.

5 Lignin: The Next Cellulose

Lignin is the second largest biopolymer after cellulose. Lignin, known as the "next cellulose," is a promising biomaterial. Lignin is emerging as a formidable biomaterial with numerous potential applications, just as cellulose has been a cornerstone of various industries due to its versatility and utility. While cellulose has long been the centre of attention, lignin is now taking its place, providing distinct properties and long-term solutions to changing environmental and industrial demands. Lignin is an important component of bioplastics because it offers a sustainable alternative to petroleum-based plastics. Because of their low toxicity, biodegradability, and environmental friendliness, lignin-based adhesives are gaining popularity in industries such as construction and woodworking. Lignin has also shown significant promise in the field of biofuels, particularly in the development of second-generation biofuels. Carbon fibres made from lignin are making a name for themselves with lightweight, high-strength materials applications. Furthermore, the antioxidant properties of lignin make it a viable candidate for use in pharmaceuticals and cosmetics.

Lignin, a complex biopolymer, is naturally found in the plant cell walls. It is extracted from pulpwood through pulping processes, including methods like kraft and sulfite. These processes yield two primary products: wood pulp (cellulose) and technical lignin. Technical lignin contains carbohydrates, inorganic substances, and extractive components. Because of its higher molecular weight, impurities, and lower reactivity, technical lignin exhibits limited performance in terms of matrix shear strength. In a study by [29], plywood was developed using lignin phenol formaldehyde (LPF) adhesive, with phenol formaldehyde adhesive as the control. The matrix shear strength test results revealed that the LPF resins exhibited lower strength than the control. This outcome can be attributed to the limited reactivity of lignin with formaldehyde, as the phenolic groups are not fully cross-linked with formaldehyde.

To realize its full potential, further purification becomes imperative. Several techniques have been devised to extract lignin from technical lignin, with processes like Ligno-Boost and macro/micro/ultrafiltration commonly employed for lignin isolation. Among these methods, the Ligno-Boost treatment is widely accepted, utilizing CO_2 and H_2SO_4 to precipitate lignin from the solution. The resulting precipitated lignin is then separated from the technical lignin through a filtration process, referred to as either purified or isolated lignin. The characteristics of purified or isolated lignin, including reactivity, viscosity, molecular weight, and the composition of monolignols, vary depending on the source and production process. The heterogeneity in these technical lignin attributes may present challenges when considering practical applications.

Purified lignin is less reactive than common monomers used in resins. This occurs because of steric hindrance and additional substituents on potential reactive sites and the phenolic hydroxy group. From an economic perspective, the practical application of purified lignin often faces challenges regarding cost-effectiveness. Consequently, additional improvements are made to isolated lignin through chemical modification. Various methods are available for lignin modification, including oxidation, reduction, methylation, phenolation, and demethylation. Among these methods, demethylation is particularly prevalent, involving removing methyl groups from lignin. This process enhances lignin's reactivity [31].

6 Lignin-Formaldehyde Adhesive

Lignin is structurally similar to phenol and is thus a promising substitute for phenol in developing eco-friendly and low-cost wood adhesives. However, the binding properties of lignin can vary depending on its bio-origin, extraction methods, and isolation methods. Lignin is a complex molecule with a heterogeneous structure, low reactive sites, and a wide range of molecular masses, unstable performance [3]. These factors can hinder the commercial utilization of lignin as an adhesive.

6.1 Lignin Urea Formaldehyde (LUF)

Lignin can be integrated with urea-formaldehyde resin, a type of adhesive typically synthesized through a two-step process involving addition and condensation reactions. Compared to other formaldehyde-based adhesives, urea-formaldehyde resin exhibits a notable affinity for water molecules. Consequently, plywood manufactured using urea-formaldehyde adhesive tends to have lower water resistance. Additionally, this adhesive contains a substantial quantity of free formaldehyde molecules, leading to formaldehyde emissions during plywood production, which can pose health concerns. The incorporation of lignin into urea-formaldehyde resin serves to enhance water resistance and reduce formaldehyde emissions. Furthermore, the introduction of lignin contributes to the improvement of various properties in plywood, including its mechanical, physical, and thermal characteristics.

As known, lignin typically exhibits lower reactivity, leading to numerous studies focusing on lignin modification. One approach involves pre-treating lignin with 1-ethyl-3-methylimidazolium acetate ionic liquid. This treatment reduces the quantity of methoxy-groups in lignin, subsequently enhancing its reactivity. Plywood bonded with this modified lignin-urea-formaldehyde (LUF) adhesive demonstrated reduced formaldehyde emissions and water adsorption compared to those bonded with unmodified LUF adhesive or commercial urea-formaldehyde (UF) resin [77].

Another method to enhance lignin reactivity is through pretreatment with methyl groups, a process known as hydroxy methylation. This modification increases the functional groups in lignin. Adding maleic anhydride-treated lignin to a UF resin has shown promising results in terms of performance. Additionally, the incorporation of methylolated sodium lignosulfonate has been found to extend the shelf life of UF resin [19]. While chemical modifications such as hydroxy methylation and ionic-liquid treatments can increase lignin reactivity, it is important to note that these lignin activation reagents are derived from fossil resources, posing environmental and sustainability issues. As a result, the primary goal of lignin-based urea-formaldehyde resin research is to develop environmentally friendly, efficient, and sustainable lignin activation methods.

6.2 Lignin-Phenol-Formaldehyde (LPF)

LPF adhesive is a type of adhesive that exhibits a wide range of properties like high bonding strength, water resistance, heat resistance, and chemical stability, making them suitable for wood processing industries such as plywood, particleboard, and fiber boards [4, 20]. LPF adhesive is a promising alternative to traditional phenol-formaldehyde adhesives, offering improved sustainability and utilizing lignin as a renewable resource [17, 75]. However, LPF adhesives have a higher viscosity, are more deeply coloured, and require more severe curing conditions than UF and PF adhesives, making them unattractive to plywood production. Modifying lignin is one way to mitigate these issues but increase the production cost of LPF adhesives, which cannot compete in the production cost of UF, which dominates the global wood adhesives market, and in performance with PF resins for preparing plywood.

The preparation of LPF adhesives via modification of lignin has come a long way, and in the last decade, various methods have been developed. These methods can be broadly classified into chemical, physical, and biological modifications. Due to its high lignin content, the most preferred raw material is black liquor from the pulping process [55]. Chemical modifications include methods such as sulfonation, phenolation, acetylation, and demethylation, which alter the structure and properties of lignin. Physical modifications involve grinding, milling, and heat treatment, which enhance the reactivity and solubility of lignin. Biological modifications utilize enzymes or microorganisms to modify lignin, offering a more environmentally friendly approach [52].

Several trial experiments have been reported to prepare LPF adhesive by substituting various amounts of technical lignin. According to a study conducted by [29], the modulus of elasticity of plywood bonded with lignosulfonate-phenol-formaldehyde resin (Lignin-formaldehyde type) (7.60 GPa) is greater than that of plywood bonded with phenol-formaldehyde adhesive (5.3 GPa). According to the findings, lignin-phenol formaldehyde adhesive is a promising alternative to phenol-formaldehyde adhesive for plywood applications. While the other properties of the two adhesives are similar, the higher modulus of elasticity of lignin-phenol formaldehyde adhesive can improve the structural performance of plywood.

A study by [70] showed that halogen acids, such as HI and HBr, are the best demethylating agents for lignin. After modification, up to 50% of phenol can be replaced in lignin-phenol formaldehyde adhesives. The modified adhesive performed better than alkaline lignin phenol formaldehyde, and the plywood met the Chinese national standard for exterior use.

Zang et al. used furfural as a crosslinking agent in lignin-phenol-formaldehyde and observed that the furfural-modified lignin-phenol-formaldehyde (LPFF) adhesive with 15% furfural content and 50% lignin substituent exhibited outstanding properties. Compared to LPF, LPFF has a high penetration ability into the wood porous structure because of lower viscosity and the higher solid content, which enable stronger adhesion. The properties of LPF adhesives can be optimized by adjusting parameters such as lignin concentration, formaldehyde to phenol molar ratio, catalyst concentration, reaction time, and temperature during the preparation process [81].

6.3 Lignin-Melamine-Formaldehyde (LMF)

Lignin-melamine-formaldehyde (LMF) is a type of lignin-based adhesive in which lignin, melamine, and formaldehyde are mixed in a constant ratio. It is used for preparing plastic materials, re-tanning leathers, coatings, biorefinery processes, and wood adhesives. However, LMF-based plywood studies still need to be well-explored. The high production expenses associated with melamine-formaldehyde resins can be attributed to the significant expense of melamine. Moreover, these resins exhibit relatively poor toughness, constraining their practical applications. Consequently, a recent shift has been towards incorporating lignin to enhance and modify melamine-formaldehyde resins. This is achieved by a three-step modification process: tosylation of lignin, synthesis of lignin-melamine (LM) copolymer, and formation of methylol LM [15].

7 Lignin-Formaldehyde-Free Adhesive

The use of formaldehyde in conventional adhesives has been a major safety concern for industry and consumers. In recent years, there have been many efforts to develop formaldehyde-free adhesives that are safe and environmentally friendly. Lignin-based adhesives have been shown to have comparable performance to formaldehyde-based adhesives, and they offer the added benefits of being non-toxic and biodegradable. Some of the most promising lignin-based adhesives include lignin-furfural [16, 80], lignin-polyurethane [9, 18, 40, 48], lignin-soy protein [10, 56, 57, 72], lignin-starch [23, 49], lignin-glyoxal [66, 68], and lignin-tannin [50, 59].

One important thing is that the bonding strength of plywood bonded with lignin deepens on the condensation degrees of lignin. As known, lignin has inherent adhesive properties, which are reduced during extraction. Extraction at high temperatures and chemicals leads to a high degree of condensation of lignin, which limits the further reaction of lignin during plywood making. Production of uncondensed or limited condensed lignin will be the future wood adhesive; it will work itself as a good wood adhesive without any treatment. Yang et al. [76] performed by using uncondensed lignin as a bio-wood adhesive to evaluate plywood's technological characteristics. Eucalyptus wood lignins are prepared using different methods, including acetone-protected lignin (KPL), Formaldehyde-Protected Lignin (FPL), Milled Wood Lignin (MWL), Kraft-Lignin (KL), Dioxane–HCl Lignin (DL) and Deep Eutectic Solvent-extracted Lignin (DESL). Results revealed that slightly condensed or protected lignins from different sources could be directly used as wood adhesives without additional physical or chemical treatments.

7.1 Lignin-Furfural (LFF)

Furfural is a bio-origin compound produced by the dehydration of sugars from agricultural byproducts. Furfural can be substituted for formaldehyde easily because of the similar reactive tendency of aromatic compounds and other aldehydes [46]. Hydrolysis lignin is usually used for the preparation of lignin-furfural adhesive. In this, Lignin-lignin and lignin-furfural condensation are the major reaction mechanisms. A maximum of 16% furfural can be blended with unmodified lignin for the best adhesive performance (Tensile strength 9.96 GPa). However, this adhesive is still inferior to PF tensile strength [16]. However, lignin-furfural synthesized with modified lignin with phenol (phenolation) and furfural shows higher tensile strength than PF resin [79]. High curing temperature and low curing rate of lignin-furfural adhesive make them unattractive to plywood industries, and pre-activation of lignin for better reaction requires petroleum-based reagents that increase the production cost and pose environmental hazards. Therefore, optimizing the adhesive formulation and avoiding toxic petro-based chemicals for green adhesive production is necessary.

7.2 Lignin-Polyurethane (LPU)

Polyurethanes (PUs) are versatile polymers with many properties and applications. PU exhibits remarkable qualities, including high resilience, strong bonding capabilities, and fatigue resistance, making it a sought-after choice for wood adhesives. Nonetheless, the inherent drawbacks of polyurethane are related to contributions to environmental pollution, limited biodegradability, and high production expenses. PU resins represent one of the most prevalent choices among cold-set adhesives. PU resins, characterized by their isocyanate base, are compounds associated with notable hazards. Applying PU resins typically necessitates adequate ventilation and high-quality personal protective equipment. Nevertheless, it is imperative to acknowledge that PU resins release dangerous gases (hydrogen cyanide). There is a compelling case for developing formaldehyde-free and non-isocyanate cold-setting wood adhesives in this context.

Recent research has explored the potential of Lignin-Based non-isocyanate Polyurethane (L-NIPU) Adhesives for wood composites, as rigorously assessed by [27]. L-NIPUs offer improved mechanical properties, lower curing temperatures, shorter pressing durations, and complete elimination of isocyanates. Lignin plays a crucial role within the Polyurethane (PU) matrix, effectively preventing plywood delamination while enhancing thermal stability and abrasion [30, 65, 67]. Additionally, L-NIPUs exhibit reduced setting times, superior wooden substrate adhesion, and extended setting periods, resulting in higher shear strength [78]. However, it's worth noting that lignin content exceeding 60% leads to increased adhesives' brittleness and delamination issues [61]. Incorporating lignin's aromatic rings significantly enhances film toughness. A clear trend in the thermal properties of L-NIPUs shows a positive correlation between lignin concentration and both the system's molecular weight and glass transition temperature [27].

Lignin, a natural polymer abundant in functional groups, has gained attention as a potential renewable raw material for lignin-based PU synthesis to reduce the pressure on renewable resources for PU synthesis [39] and free isocyanate content [18]. Furthermore, LPU adhesives offer a cost-effective alternative to traditional polyurethane wood adhesives [40], largely due to the economical nature of lignin. Lignin boasts phenolic and aliphatic hydroxyl groups, serving as potential polyols that can react with isocyanates to yield an LPU adhesive [71]. LPU adhesives can be synthesized by direct lignin incorporation, chemical lignin modification, and depolymerization [21].

The extensive adoption of PU resin adhesive within the wood-based composites industry can be attributed to its outstanding attributes. Plywood bonded with PU resins has demonstrated dry tensile shear strength (TSS) values within the 1.2–2.5 MPa range [42]. Several research studies have evaluated the feasibility of incorporating lignin with PU and its adhesive characterization, but the typical LPU-based plywood was not explored well [6, 9, 12, 13, 18, 35, 38, 48, 51, 64]. Chauhan et al. [9] modified lignin with methylene diphenyl diisocyanate (MDI) and prepared an LPU adhesive by reacting modified lignin with isocyanate. The formulated adhesive had excellent water resistance and showed a dry wood panel strength of 13.5 MPa and a wet strength of 9.39 MPa. The glass transition temperature increased with the addition of lignin.

Nacas et al. [48] synthesized LPU wood adhesives using technical kraft lignin and methylene diphenyl diisocyanate (MDI). The reaction scheme involved mechanical mixing of lignin and MDI, with varying NCO: OH ratios explored, and the highest shear strength achieved in their experiments was 6.80 MPa, indicating the adhesive's effectiveness for bonding woody surfaces.

Nacas et al. [18] conducted a study to assess the mechanical, thermal properties, and chemical structure of Lignin-Polyurethane (LPU) adhesives. This study showed that as the weight percentage of lignin increased, there was a noticeable decrease in the free isocyanate content, resulting in a longer setting time. Concurrently, the increase in lignin content was associated with higher shear strength values. Furthermore, increased lignin content correlated with increased glass transition temperature.

In light of the relatively low OH content inherent in lignin, it is often imperative to undertake chemical modifications to ensure that the resulting adhesive meets the requisite performance standards. A prominent approach to enhancing the bonding strength of the adhesive involves an increased utilization of isocyanate. However, it is important to acknowledge that this augmentation in isocyanate usage also increases production costs. To optimize the cost-effectiveness of lignin-polyurethane, forthcoming research endeavors should prioritize strategies that mitigate the dependence on isocyanate during the preparation process. This could reduce production costs and contribute to a more sustainable and environmentally friendly approach to adhesive formulation.

7.3 Lignin-Soy Protein (LSP)

Soybean protein is a favored choice for plywood adhesive formulation due to its inherent advantages: stability, cost-effectiveness, ready accessibility, and environmentally friendly nature [44]. Soy protein adhesives have garnered increasing attention as a viable alternative to formaldehyde-based resins [73]. Nevertheless, soy protein-based adhesives exhibit several challenges: a lack of a well-defined melting point, high susceptibility to water, and a tendency toward microbial growth. In response to these challenges, the incorporation of lignin has emerged as a promising solution [41].

Numerous investigations have been dedicated to addressing the issues of poor water resistance, susceptibility to microbial degradation, and suboptimal mechanical properties in soy protein adhesives. The most effective approach for enhancing these properties involves modifying these adhesives by incorporating crosslinking agents. Several methods have been explored to counteract the biological degradation of soy-based adhesives. One such method involves the treatment with preservatives, including chlorophenols, nanoparticles, and copper sulfates [28, 37, 39]. These preservatives can extend the durability of soy-based adhesives, however, their utilization relies on non-renewable fossil resources and has the drawback of contributing to heavy metal pollution.

An alternative approach involves harnessing natural resources with inherent antifungal properties, such as lignin, chitosan, and essential oils [7, 62, 74]. Among these, lignin is particularly noteworthy as it is an abundant biopolymer known for its antimicrobial attributes. The polyphenols present in lignin not only exhibit antimicrobial properties but also possess antioxidant and anti-inflammatory activities. Numerous studies have shown that incorporating lignin into soy-based wood adhesive formulations is effective. Chen et al. [10] investigated the adhesive's bonding strength and fungal growth resistance under high-humidity conditions. This research underscores the potential of incorporating lignin into soy-based adhesive formulations to enhance their antifungal and mechanical attributes, offering broader application possibilities in plywood industries. The results showed that lignin significantly improved the antifungal properties of the adhesive for interior-use plywood. This improvement is due to lignin's natural properties, particularly its phenolic hydroxyl groups and strong network structure, which prevent fungal degradation. Adding lignin also significantly increased the adhesive strength, resulting in an impressive improvement that exceeded 1.0 MPa. This represents a significant 363.60% improvement over the unmodified soy-based adhesive.

A study by Pradyawonga et al. [56] found that adhesives with smaller lignin particles had greater wet strength. Expanding on this, Pradyawonga et al. [57] introduced modified lignin to enhance the water resistance of soy protein adhesives. The resulting soy protein-lignin adhesive exhibited increased strength, improved thermal stability, and better adhesion in wet conditions. Xiao et al. [72] investigated the effects of incorporating lignin from grain sorghum stalk in two different forms: isolated sorghum lignin (ISL) and extruded sorghum lignin (ESL) into two distinct soy protein adhesives. These soy protein adhesives were soy protein isolates (SPI) and modified soy protein (MSP) achieved through chemical modification. The addition of lignin resulted in notable enhancements in the shear strength and water resistance of the LSP. Notably, all LSP formulations exhibited 100% cohesive wood failure.

Comparatively, the pure SP adhesive exhibited significantly lower shear strength. However, the introduction of lignin led to a substantial increase in shear strength. Specifically, the blended adhesive achieved 100% cohesive wood failure when the lignin content exceeded 20% for ISL or 10% for ESL. This improvement was also reflected in increased water resistance as the percentage of lignin in the blend increased. The study concluded that blending lignin could significantly enhance both the shear strength and water resistance of LSP, underscoring the potential of lignin as an effective additive for improving the performance of soy protein adhesives.

In summary, incorporating lignin into soy-based adhesives shows promise for developing future plywood adhesives. With a focus on lignin's phenolic attributes, studies have shown significant enhancements in adhesive performance, highlighting the promising role of lignin in advancing the application of soy protein adhesives in plywood industries. The synergy between soy proteins and lignin opens new doors for environmentally friendly and high-performance adhesives in the wood industry.

7.4 Lignin-Starch (LS)

Starch is one of the most abundant polysaccharides and is obtainable from various plant sources. In addition to its use in the food industry, technical starches find applications in the pulp and paper, textile, and adhesive industries [2]. Starch holds potential as a biopolymer for wood adhesives. However, due to its low reactivity and limited water resistance, its bonding capacity falls short of being effective for wood glueing. Thus, starch requires modification or grafting with suitable crosslinkers. Typically, crosslinkers such as MDI, epoxies, chitosan, and tannin are used, but the use of lignin for modifying starch adhesive for plywood still needs to be explored [14].

Starch is a particularly attractive bio-adhesive with a long history of use in the paper and corrugated board industry [49]. Recent research studies have suggested that modified starch from various sources is the most promising candidate for composite adhesives in applications like particleboards [25, 45], medium-density fiberboard [1, 43], and plywood [11, 47]. However, it is worth noting that research into starch-lignin combination wood adhesives for plywood still needs to be improved.

Hellmayr et al. [23] utilized equal quantities of corn starch and sodium lignosulfonate to investigate their potential as an alternative adhesive for heat-bonding wood. This research offers valuable insights into developing plywood that is not only environmentally responsible but also maintains competitive mechanical performance. In comparative automated bonding evaluation system (ABES) shear tests, veneers subjected to the corn starch-sodium lignosulfonate adhesive, and a commercial urea-formaldehyde resin exhibited nearly identical shear strength, both achieving approximately 8.5 N/mm^2.

Further analysis of the bonding properties revealed that starch primarily contributes to the adhesive's bonding strength. At the same time, lignin enhances the even distribution of the adhesive through its dispersing and water-retarding properties. This collaborative interaction between the two components underscores the synergy between the natural adhesive properties of starch and the dispersing and water-retarding capabilities of lignosulfonate. This synergy is pivotal in achieving the impressive shear strength observed in comparative tests. Combining corn starch and sodium lignosulfonate as the bonding agent offers an eco-friendly alternative to conventional wood adhesives. This outcome is significant as it underscores the viability of starch-lignin mixtures as a practical replacement for traditional wood adhesives without compromising the structural integrity of the final product.

7.5 Lignin-Glyoxal (LG)

In recent years, glyoxal has garnered significant attention from researchers as a promising cross-linking agent and a substitute for formaldehyde. Glyoxal, the smallest biobased dialdehyde, possesses non-volatile, biodegradable, and cost-effective properties. Notably, [66] developed a biobased phenolic adhesive utilizing lignin from corn stover biorefinery and glyoxal. Compared with PF resin and a formulated LPF resin, the resulting LG resin exhibited a higher curing temperature at 167 °C. However, the rate and degree of cure were similar. Impressively, the LG resin displayed a relatively high dry adhesion strength of 3.9 MPa, with over 90% wood failure. Nevertheless, it did not pass the wet adhesion (boiling water) test. Despite this, the formulated LG resin exhibited excellent stability when exposed to room-temperature water. It remained intact even after one week of water immersion. The high dry adhesion strength positions this lignin-based, formaldehyde-free adhesive as a unique biobased glue for producing interior-grade plywood.

Another study by Esfandiyari et al. [68] delved into the potential of lignin-glyoxal resins as adhesives in the production of poplar plywood. They explored different adhesive compositions, including pure lignin, lignin-glyoxal at 15%, and lignin-glyoxal at 30%, each with varying levels of adhesive usage. The results indicated that the lignin-glyoxal 30% adhesive performed most closely to the control glue, urea-formaldehyde resin, in terms of mechanical and physical properties. Interestingly, the adhesive type and usage amount did not significantly impact the mechanical and physical properties of the plywood, except for thickness swelling. In conclusion, lignin-glyoxal combinations present a promising alternative to commercial formaldehyde adhesives. However, unlocking the full potential of lignin-glyoxal resins may require further modification and optimization of the adhesive preparation process.

7.6 Lignin-Tannin (LT)

Tannins are bio-polyphenols that have a similar structure to phenols. Tannin composition and properties like lignin are highly source-dependent and widespread. Due to their high activity, tannins are promising ingredients for wood adhesives. Typically, tannins are treated with hexamethylenetetramine or isocyanate to produce highly reactive and rapid tannin coatings. Using both lignin and tannin, derived from biomass, gives lignin-tannin adhesives an advantage over other lignin-based adhesives. There are, however, some drawbacks to consider. These adhesives have a high viscosity during preparation and application, as well as a low level of cross-linking and a short shelf life. Inadequate cross-linking also contributes to less-than-optimal mechanical performance. Furthermore, these adhesives face a water resistance challenge, which can be mitigated by incorporating isocyanates into the formula [22]. Much effort has gone into developing lignin-tannin adhesives.

Rhazi et al. [59] investigated the potential of utilizing a sustainable adhesive composition for plywood production. This eco-friendly adhesive consisted of 75% tannin extracted from the Moroccan bark of *Acacia mollissima* and 25% glyoxalated lignosulfonates. The result revealed that plywood panels bonded with this unique combination exhibited mechanical properties on par with those constructed using commercial UF and PF resins. Notably, the tensile test results demonstrated a maximum rupture strength of approximately 2950 N for the tannin-glyoxalated lignosulfonates adhesive, slightly below the 3050 N of PF but surpassing the 2750 N of UF. This research showcases the promise of sustainable, tannin-based adhesives in the woodworking industry, offering a more environmentally friendly alternative to conventional resin options.

Navarrete et al. [50] prepared three-ply triplicate plywood using 94% eco-friendly materials, including lignin and tannin as glue components of total adhesive and non-natural components like hexamine and glyoxal account for 0.5% incorporated to improve the glue performance. The plywood itself is, thus, 99.5% eco-friendly. The result showed that plywood panels exhibit good dry tensile strength, accounting for 1.2 MPa. Similarly, Lei et al. [36] investigated the properties of calcium lignosulfonate-tannin adhesives in depth. Unlike [50], varying levels of MDI (methylene diphenyl diisocyanate) were added to improve the adhesive's overall performance. The researchers discovered a significant difference in the adhesive's behaviour based on the lignin's molecular weight. Regarding internal bonding strength, adhesives made with low-molecular-weight lignin outperformed those made with high-molecular-weight lignin. Furthermore, it was discovered that the amount of MDI added to the formulation had a significant impact on the strength of the adhesive. These findings suggest that selecting lignin with a specific molecular weight can be crucial in optimizing adhesive performance. Additionally, understanding MDI dosage's influence can help fine-tune the adhesive formulation to achieve desired strength levels. Overall, all the research provides valuable insights for developing high-performance adhesives tailored to specific applications.

8 Life Cycle Assessment (LCA)

Lignin possesses inherent binding properties activated under high temperature, pressure, and prolonged pressing time. Lignin is crucial in binderless boards, which are manufactured without adhesives. The pressing time and temperature significantly influence the breakdown of hemicelluloses and lignin in binderless boards, and the strength is primarily attributed to lignin cross-linking mechanisms, particularly lignin-lignin and lignin-polysaccharide cross-linking reactions. From an economic standpoint, binderless boards necessitate greater investments in time and energy. On the other hand, synthetic adhesives are restricted due to concerns about their impact on environmental health. Given these considerations, renewable adhesives emerge as a more favorable option in the present context, given their potential to align with economic and environmental objectives. However, bio-adhesives, especially for woodworking, require additional support from cross-linkers to improve their binding ability for specific uses. Cross-linkers are compounds (whether synthetic or bio-based) used to enhance the binding capability of adhesives. During the production of primary adhesive raw materials (for instance, lignin and cross-linkers) and adhesives, there is a chance for environmental burdens to emerge.

In environmental concerns, it is critical to evaluate the environmental friendliness of bio-based adhesives. Some bio-based adhesives, particularly those derived from soy protein and tannin, have received eco-friendliness recognition and have become commercially available. All bio-based adhesives do not have the same eco-friendliness credentials. At various stages of the manufacturing process, various environmental burdens emerge during the production of bio-based adhesives. The use of chemicals and electricity, the production of wastewater, and the emission of pollutants into the atmosphere are all examples of these burdens. LCA is a rigorous and widely used environmental methodology. It is an effective tool for quantifying and evaluating the environmental impacts of a given product, process, or activity over its entire life cycle.

Lignin can be extracted using different processes with various chemicals; solvent pulping contributes to a lower environmental impact than the other processes. In the lignin extraction process by solvent pulping, zero waste stream is generated, while the kraft lignin extraction process produces liquid and gaseous waste streams. Additionally, oregano solvent lignin has the advantage of smaller particles with better dispersity, low molecular weight, and higher reactivity than lignin from other processes. Likewise, advanced biorefinery separation processes such as ionic liquid, supercritical solvent, and deep eutectic solvent mixture extraction processes are expected to perform better. However, solvent recovery and cost are major issues that limit industrial acceptance [53].

Next is the cross-linker and the environmental burdens that emerge during its production and use in adhesive preparation. Synthetic cross-linkers have a negative impact on the environment, for example, Methylolation (the addition of formaldehyde). Bio-based cross-linkers, such as glyoxal (the addition of glyoxal is known as glyoxalation), are frequently reported in the literature as an alternative to conventional methylolation. Glyoxal is a dialdehyde, while formaldehyde is a monoaldehyde compound. Glyoxal exhibits two carbonyl compounds and has been proven to be a suitable alternative candidate for formaldehyde as a cross-linker. Moreover, glyoxal is derived from renewable sources, while formaldehyde is derived from petroleum. Other suitable cross-linkers are gossypol, glutaraldehyde, and propionaldehyde, but they have issues with solubility and toxicity. Lignin-glyoxal adhesive has been proven to be the best alternative to conventional adhesive, with acceptable adhesive performance except for boiling waterproof (BWP). However, it should be noted that being "bio-based" does not necessarily guarantee an improved carbon footprint. In glyoxal production, oxidising (bioderived) acetaldehyde with nitric acid may lead to N_2O emissions, a potent greenhouse gas. This could overcompensate for CO_2 savings obtained elsewhere [69]. Arias et al. [5] used Life Cycle Assessment to evaluate the production systems of wood adhesives (soy, tannin, and lignin-based) from cradle-to-gate. Soy and tannin adhesives have an overall better environmental profile than fossil resins. Lignin-based adhesives have higher single environmental scores, which means higher impacts on the environment than conventional adhesives.

9 Conclusion

Lignin is an abundant, low-cost, and renewable material that has the potential to be used as an industrial adhesive. However, several limiting factors have prevented the widespread adoption of lignin-based adhesives, including low reactivity, complex structure, high viscosity, longer hot-pressing time, and higher temperature. Additionally, the performance of lignin-based adhesives can be inconsistent due to the variation in lignin properties with different plants and extraction methods. Despite these challenges, some progress is being made in developing lignin-based adhesives. For example, lignin phenolic resin/lignin bio-based resin has been shown to meet the requirements of some relevant standards. Lignin, whether employed independently or in conjunction with other biopolymer matrices and synthetic polymers, has exhibited significant promise in the realm of wood adhesive production. While most adhesive experimental studies are conducted at the laboratory level, their viability for industrial applications cannot be considered critically without taking into account both economic and environmental factors.

It is essential to recognize that pursuing lignin-based adhesives is just the initial step. To genuinely assess their ecological benefits, we must extend our research focus towards understanding the environmental implications of the adhesive production process, including the potential pollutants generated. Consequently, comprehensive life cycle assessments (LCAs) are imperative, delving beyond traditional analyses of adhesive physical, mechanical, thermal, chemical, and morphological properties. In light of the low reactivity of lignin, numerous approaches have been developed to enhance its utility. However, the optimization of all parameters, including chemical concentrations, energy consumption, and pH, is crucial to achieve the best outcomes.

Recent studies have explored diverse combinations of lignin with other polymer matrices for plywood adhesive applications. Consequently, future research endeavors should emphasize the optimization of plywood adhesive preparation methodologies and the discovery of safe, effective, and environmentally friendly lignin extraction, purification, and modification techniques. These key considerations are pivotal to advancing the utilization of lignin and other biocompatible materials for sustainable wood adhesive solutions.

References

1. Abbott, A.P., Palazuela Conde, J., Davis, S.J., Wise, W.R.: Starch as a replacement for urea-formaldehyde in medium density fibreboard. Green Chem. **14**, 3067 (2012). https://doi.org/10.1039/c2gc36194a
2. Adewale, P., Yancheshmeh, M.S., Lam, E.: Starch modification for non-food, industrial applications: market intelligence and critical review. Carbohyd. Polym. **291**, 119590 (2022). https://doi.org/10.1016/j.carbpol.2022.119590
3. Ang, A.F., Ashaari, Z., Lee, S.H., et al.: Lignin-based copolymer adhesives for composite wood panels – a review. Int. J. Adhes. Adhes. **95**, 102408 (2019). https://doi.org/10.1016/j.ijadhadh.2019.102408
4. Antov, P., Savov, V., Mantanis, G.I., Neykov, N.: Medium-density fibreboards bonded with phenol-formaldehyde resin and calcium lignosulfonate as an eco-friendly additive. Wood Mat. Sci. Eng. **16**, 42–48 (2020). https://doi.org/10.1080/17480272.2020.1751279
5. Arias, A., González-García, S., González-Rodríguez, S., et al.: Cradle-to-gate life cycle assessment of bio-adhesives for the wood panel industry a comparison with petrochemical alternatives. Sci. Total. Environ. **738**, 140357 (2020). https://doi.org/10.1016/j.scitotenv.2020.140357
6. Avelino, F., Almeida, S.L., Duarte, E.B., et al.: Thermal and mechanical properties of coconut shell lignin-based polyurethanes synthesized by solvent-free polymerization. J. Mater. Sci. **53**, 1470–1486 (2017). https://doi.org/10.1007/s10853-017-1562-z
7. Azadfar, M., Gao, A.H., Bule, M.V., Chen, S.: Structural characterization of lignin: a potential source of antioxidants guaiacol and 4-vinylguaiacol. Int. J. Biol. Macromol. **75**, 58–66 (2015). https://doi.org/10.1016/j.ijbiomac.2014.12.049
8. Burke, D.: Sitting pretty: the history and restoration of antique chairs. Fine Art Restoration Company (2022). https://fineart-restoration.co.uk/news/sitting-pretty-the-history-and-restoration-of-antique-chairs/
9. Chauhan, M., Gupta, M., Singh, B., et al.: Effect of functionalized lignin on the properties of lignin–isocyanate prepolymer blends and composites. Eur. Polym. J. **52**, 32–43 (2014). https://doi.org/10.1016/j.eurpolymj.2013.12.016
10. Chen, S., Chen, H., Yang, S., Fan, D.: Developing an antifungal and high-strength soy protein-based adhesive modified by lignin-based polymer. Ind. Crops Prod. **170**, 113795 (2021). https://doi.org/10.1016/j.indcrop.2021.113795
11. Chen, X., Sun, C., Wang, Q., et al.: Preparation of glycidyl methacrylate grafted starch adhesive to apply in high-performance and environment-friendly plywood. Int. J. Biol. Macromol. **194**, 954–961 (2022). https://doi.org/10.1016/j.ijbiomac.2021.11.152
12. Chen, Y., Fu, S., Zhang, H.: Signally improvement of polyurethane adhesive with hydroxy-enriched lignin from bagasse. Colloids Surf. A **585**, 124164 (2020). https://doi.org/10.1016/j.colsurfa.2019.124164
13. de Haro, J.C., Allegretti, C., Smit, A.T., et al.: Biobased polyurethane coatings with high biomass content: tailored properties by lignin selection. ACS Sustain. Chem. & Eng. **7**, 11700–11711 (2019). https://doi.org/10.1021/acssuschemeng.9b01873

14. Din, Z., Chen, L., Xiong, H., et al.: Starch: an undisputed potential candidate and sustainable resource for the development of wood adhesive. Starch - Stärke **72**, 1900276 (2020). https://doi.org/10.1002/star.201900276
15. Diop, A., Adjallé, K., Boëns, B., et al.: Synthesis and characterization of lignin–melamine–formaldehyde resin. J. Thermoplast. Compos. Mater. **30**, 1255–1266 (2016). https://doi.org/10.1177/0892705716632856
16. Dongre, P., Driscoll, M., Amidon, T., Bujanovic, B.: Lignin-furfural based adhesives. Energies **8**, 7897–7914 (2015). https://doi.org/10.3390/en8087897
17. Ferdosian, F., Pan, Z., Gao, G., Zhao, B.: Bio-based adhesives and evaluation for wood composites application. Polymers **9**, 70 (2017). https://doi.org/10.3390/polym9020070
18. Gadhave, R.V., Kasbe, P.S., Mahanwar, P.A., Gadekar, P.T.: Synthesis and characterization of lignin-polyurethane based wood adhesive. Int. J. Adhes. Adhes. **95**, 102427 (2019). https://doi.org/10.1016/j.ijadhadh.2019.102427
19. Gao, S., Cheng, Z., Zhou, X., et al.: Unexpected role of amphiphilic lignosulfonate to improve the storage stability of urea formaldehyde resin and its application as adhesives. Int. J. Biol. Macromol. **161**, 755–762 (2020). https://doi.org/10.1016/j.ijbiomac.2020.06.135
20. Ghorbani, M., Liebner, F., Van Herwijnen, H.W.G., et al.: Lignin phenol formaldehyde resoles: the impact of lignin type on adhesive properties. BioResources **11**, 6727–6741 (2016). https://doi.org/10.15376/biores.11.3.6727-6741
21. Gouveia, J.R., da Costa, C.L., Tavares, L.B., dos Santos, D.J.: Synthesis of lignin-based polyurethanes: a mini-review. Mini-Rev. Org. Chem. **16**, 345–352 (2019). https://doi.org/10.2174/1570193x15666180514125817
22. Hage, R.E., Brosse, N., Navarrete, P., Pizzi, A.: Extraction, characterization and utilization of organosolv miscanthus lignin for the conception of environmentally friendly mixed tannin/lignin wood resins. J. Adhes. Sci. Technol. **25**, 1549–1560 (2011). https://doi.org/10.1163/016942410x524110
23. Hellmayr, R., Šernek, M., Myna, R., et al.: Heat bonding of wood with starch-lignin mixtures creates new recycling opportunities. Mater. Today Sustain. **19**, 100194 (2022). https://doi.org/10.1016/j.mtsust.2022.100194
24. Huang, C., Peng, Z., Li, J., et al.: Unlocking the role of lignin for preparing the lignin-based wood adhesive: a review. Ind. Crops Prod. **187**, 115388 (2022). https://doi.org/10.1016/j.indcrop.2022.115388
25. Huang, H.-K., Hsu, C.-H., Hsu, P.-K., et al.: Preparation and evaluation of particleboard from insect rearing residue and rice husks using starch/citric acid mixture as a natural binder. Biomass Convers. Biorefin. **12**, 633–641 (2020). https://doi.org/10.1007/s13399-020-00994-6
26. Hussin, M.H., Abd Latif, N.H., Hamidon, T.S., et al.: Latest advancements in high-performance bio-based wood adhesives: a critical review. J. Market. Res. **21**, 3909–3946 (2022). https://doi.org/10.1016/j.jmrt.2022.10.156
27. Iswanto, A.H., Lubis, M.A.R., Sutiawan, J., et al.: Latest advancements in the development of high-performance lignin- and tannin-based non-isocyanate polyurethane adhesive for wood composites. Polymers **15**, 3864 (2023). https://doi.org/10.3390/polym15193864
28. Jin, S., Li, K., Gao, Q., et al.: Multiple crosslinking strategy to achieve high bonding strength and antibacterial properties of double-network soy adhesive. J. Clean. Prod. **254**, 120143 (2020). https://doi.org/10.1016/j.jclepro.2020.120143
29. Karri, R., Lappalainen, R., Tomppo, L., Yadav, R.: Bond quality of poplar plywood reinforced with hemp fibers and lignin-phenolic adhesives. Compos. Part C: Open Access **9**, 100299 (2022). https://doi.org/10.1016/j.jcomc.2022.100299
30. Kazemi Najafi, S., Najafi, F., Pizzi, A., et al.: Effect of amine type on lignin modification to evaluate its reactivity in polyol construction for Non-Isocyanate Polyurethanes (NIPU). J. Renew. Mater. **11**, 2171–2189 (2023). https://doi.org/10.32604/jrm.2023.027835
31. Kienberger, M., Maitz, S., Pichler, T., Demmelmayer, P.: Systematic review on isolation processes for technical lignin. Processes **9**, 804 (2021). https://doi.org/10.3390/pr9050804
32. Korányi, T.I., Fridrich, B., Pineda, A., Barta, K.: Development of 'lignin-first' approaches for the valorization of lignocellulosic biomass. Molecules **25**, 2815 (2020). https://doi.org/10.3390/molecules25122815

33. Kozowyk, P.R.B., van Gijn, A.L., Langejans, G.H.J.: Understanding preservation and identification biases of ancient adhesives through experimentation. Archaeol. Anthropol. Sci. **12**, 1–17 (2020). https://doi.org/10.1007/s12520-020-01179-y
34. Kumar, C., Leggate, W.: An overview of bio-adhesives for engineered wood products. Int. J. Adhes. Adhes. **118**, 103187 (2022). https://doi.org/10.1016/j.ijadhadh.2022.103187
35. Lee, A., Deng, Y.: Green polyurethane from lignin and soybean oil through non-isocyanate reactions. Eur. Polym. J. **63**, 67–73 (2015). https://doi.org/10.1016/j.eurpolymj.2014.11.023
36. Lei, H., Pizzi, A., Du, G.: Environmentally friendly mixed tannin/lignin wood resins. J. Appl. Polym. Sci. **107**, 203–209 (2007). https://doi.org/10.1002/app.27011
37. Li, H., Liang, Y., Li, P., He, C.: Conversion of biomass lignin to high-value polyurethane: a review. J. Bioresour. Bioprod. **5**, 163–179 (2020). https://doi.org/10.1016/j.jobab.2020.07.002
38. Li, H., Sun, J.-T., Wang, C., et al.: High modulus, strength, and toughness polyurethane elastomer based on unmodified lignin. ACS Sustain. Chem. & Eng. **5**, 7942–7949 (2017). https://doi.org/10.1021/acssuschemeng.7b01481
39. Li, W., Chen, M., Li, Y., et al.: Improving mildew resistance of soy meal by nano-Ag/TiO2, zinc pyrithione and 4-cumylphenol. Polymers **12**, 169 (2020). https://doi.org/10.3390/polym12010169
40. Lima García, J., Pans, G., Phanopoulos, C.: Use of lignin in polyurethane-based structural wood adhesives. J. Adhes. **94**, 814–828 (2018). https://doi.org/10.1080/00218464.2017.1385458
41. Lizundia, E., Sipponen, M.H., Greca, L.G., et al.: Multifunctional lignin-based nanocomposites and nanohybrids. Green Chem. **23**, 6698–6760 (2021). https://doi.org/10.1039/d1gc01684a
42. Lubis, M.A.R., Labib, A., Sudarmanto, et al.: Influence of lignin content and pressing time on plywood properties bonded with cold-setting adhesive based on poly (vinyl alcohol), lignin, and hexamine. Polymers **14**, 2111 (2022). https://doi.org/10.3390/polym14102111
43. Lubis, M.A.R., Park, B.-D., Kim, Y.-S., et al.: Visual inspection of surface mold growth on medium-density fiberboard bonded with oxidized starch adhesives. Wood Mat. Sci. Eng. **18**, 819–826 (2022). https://doi.org/10.1080/17480272.2022.2073828
44. Ma, Y., Kou, Z., Hu, Y., et al.: Research advances in bio-based adhesives. Int. J. Adhes. Adhes. **126**, 103444 (2023). https://doi.org/10.1016/j.ijadhadh.2023.103444
45. Mensah, P., Mitchual, S.J., Apetorgbor, M.M., Frimpong-Mensah, K.: Decay resistance of particleboards manufactured with four agro-forest residues using cassava starch and urea formaldehyde as adhesives. J. Agric. Chem. Environ. **09**, 240–254 (2020). https://doi.org/10.4236/jacen.2020.94019
46. Mittal, A., Black, S.K., Vinzant, T.B., O'Brien, M., Tucker, M.P., Johnson, D.K.: Production of furfural from process-relevant biomass-derived pentoses in a biphasic reaction system. ACS Sustain. Chem. Eng. **5**, 5694–5701 (2017). https://doi.org/10.1021/acssuschemeng.7b00215
47. Moubarik, A., Allal, A., Pizzi, A., et al.: Characterization of a formaldehyde-free cornstarch-tannin wood adhesive for interior plywood. Eur. J. Wood Wood Prod. **68**, 427–433 (2009). https://doi.org/10.1007/s00107-009-0379-0
48. Nacas, A.M., Ito, N.M., Sousa, R.R.D., et al.: Effects of NCO:OH ratio on the mechanical properties and chemical structure of Kraft lignin–based polyurethane adhesive. J. Adhes. **93**, 18–29 (2016). https://doi.org/10.1080/00218464.2016.1177793
49. Nasiri, A., Wearing, J., Dubé, M.A.: Using lignin to modify starch-based adhesive performance. Chem. Eng. **4**, 3 (2020). https://doi.org/10.3390/chemengineering4010003
50. Navarrete, P., Mansouri, H.R., Pizzi, A., et al.: Wood panel adhesives from low molecular mass lignin and tannin without synthetic resins. J. Adhes. Sci. Technol. **24**, 1597–1610 (2010). https://doi.org/10.1163/016942410x500972
51. de Oliveira, F., Ramires, E.C., Frollini, E., Belgacem, M.N.: Lignopolyurethanic materials based on oxypropylated sodium lignosulfonate and castor oil blends. Ind. Crops Prod. **72**, 77–86 (2015). https://doi.org/10.1016/j.indcrop.2015.01.023
52. Peng, Z., Jiang, X., Si, C., et al.: Advances of modified lignin as substitute to develop lignin-based phenol-formaldehyde resin adhesives. Chemsuschem **16**, e202300174 (2023). https://doi.org/10.1002/cssc.202300174

53. Perederic, O.A., Mountraki, A., Papadopoulou, E., et al.: Life cycle analysis of phenol - formaldehyde resins substituted with lignin. Comput. Aided Chem. Eng. **48**, 607–612 (2020). https://doi.org/10.1016/b978-0-12-823377-1.50102-6
54. Plyreporter: IPIRTI seeks BIS standards for formaldehyde emission from wood based panel products (2016). https://www.plyreporter.com/article/71631/ipirti-seeks-bis-standards-for-formaldehyde-emission-from-wood-based-panel-products
55. Pola, L., Collado, S., Oulego, P., Díaz, M.: Kraft black liquor as a renewable source of value-added chemicals. Chem. Eng. J. **448**, 137728 (2022). https://doi.org/10.1016/j.cej.2022.137728
56. Pradyawong, S., Qi, G., Li, N., et al.: Adhesion properties of soy protein adhesives enhanced by biomass lignin. Int. J. Adhes. Adhes. **75**, 66–73 (2017). https://doi.org/10.1016/j.ijadhadh.2017.02.017
57. Pradyawong, S., Qi, G., Sun, X.S., Wang, D.: Laccase/TEMPO-modified lignin improved soy-protein-based adhesives: adhesion performance and properties. Int. J. Adhes. Adhes. **91**, 116–122 (2019). https://doi.org/10.1016/j.ijadhadh.2019.03.005
58. Raydan, N.D.V., Leroyer, L., Charrier, B., Robles, E.: Recent advances on the development of protein-based adhesives for wood composite materials—a review. Molecules **26**, 7617 (2021). https://doi.org/10.3390/molecules26247617
59. Rhazi, N., Oumam, M., Sesbou, A., et al.: Physico-mechanical properties of plywood bonded with ecological adhesives from Acacia mollissima tannins and lignosulfonates. Eur. Phys. J. Appl. Phys. **78**, 34813 (2017). https://doi.org/10.1051/epjap/2017170067
60. Risholm-Sundman, M., Larsen, A., Vestin, E., Weibull, A.: Formaldehyde emission—comparison of different standard methods. Atmos. Environ. **41**, 3193–3202 (2007). https://doi.org/10.1016/j.atmosenv.2006.10.079
61. Saito, T., Perkins, J.H., Jackson, D.C., et al.: Development of lignin-based polyurethane thermoplastics. RSC Adv. **3**, 21832 (2013). https://doi.org/10.1039/c3ra44794d
62. Salem, M.Z.M., Zidan, Y.E., Mansour, M.M.A., et al.: Antifungal activities of two essential oils used in the treatment of three commercial woods deteriorated by five common mold fungi. Int. Biodeterior. Biodegrad. **106**, 88–96 (2016). https://doi.org/10.1016/j.ibiod.2015.10.010
63. Salthammer, T.: The formaldehyde dilemma. Int. J. Hyg. Environ. Health **218**, 433–436 (2015). https://doi.org/10.1016/j.ijheh.2015.02.005
64. Sažin, J., Pizzi, A., Amirou, S., et al.: Organosolv lignin for Non-Isocyanate Based Polyurethanes (NIPU) as wood adhesive. J. Renew. Mater. **9**, 881–907 (2021). https://doi.org/10.32604/jrm.2021.015047
65. Sažin, J., Poljanšek, I., Pizzi, A., Šernek, M.: Curing kinetics of tannin and lignin biobased adhesives determined by DSC and ABES. J. Renew. Mater. **10**, 2117–2131 (2022). https://doi.org/10.32604/jrm.2022.019602
66. Siahkamari, M., Emmanuel, S., Hodge, D.B., Nejad, M.: Lignin-glyoxal: a fully biobased formaldehyde-free wood adhesive for interior engineered wood products. ACS Sustain. Chem. & Eng. **10**, 3430–3441 (2022). https://doi.org/10.1021/acssuschemeng.1c06843
67. Sternberg, J., Pilla, S.: Materials for the biorefinery: high bio-content, shape memory kraft lignin-derived non-isocyanate polyurethane foams using a non-toxic protocol. Green Chem. **22**, 6922–6935 (2020). https://doi.org/10.1039/d0gc01659d
68. Tupa Esfandiyari, M.R., Talaei Pour, M., Khademieslam, H., et al.: Investigating the possibility of making lignin-glyoxal resins as adhesives in the production of plywood. BioResources **14**, 7122–7133 (2019). https://doi.org/10.15376/biores.14.3.7122-7133
69. Van Nieuwenhove, I., Renders, T., Lauwaert, J., et al.: Biobased resins using lignin and glyoxal. ACS Sustain. Chem. & Eng. **8**, 18789–18809 (2020). https://doi.org/10.1021/acssuschemeng.0c07227
70. Wang, H., Eberhardt, T.L., Wang, C., et al.: Demethylation of alkali lignin with halogen acids and its application to phenolic resins. Polymers **11**, 1771 (2019). https://doi.org/10.3390/polym11111771
71. Wu, M., Peng, J., Dong, Y., et al.: Extraction and oxypropylation of lignin by an efficient and mild integration process from agricultural waste. Ind. Crops Prod. **172**, 114013 (2021). https://doi.org/10.1016/j.indcrop.2021.114013

72. Xiao, Z., Li, Y., Wu, X., et al.: Utilization of sorghum lignin to improve adhesion strength of soy protein adhesives on wood veneer. Ind. Crops Prod. **50**, 501–509 (2013). https://doi.org/10.1016/j.indcrop.2013.07.057
73. Xie, D.-Y., Song, F., Zhang, M., et al.: Roles of soft segment length in structure and property of soy protein isolate/waterborne polyurethane blend films. Ind. Eng. Chem. Res. **55**, 1229–1235 (2016). https://doi.org/10.1021/acs.iecr.5b04185
74. Xing, F., Chen, H., Zhang, S., et al.: Effect of p-cumylphenol on the mold resistance of modified soybean flour adhesive and poplar plywood. BioResources **10**, 1543–1552 (2015). https://doi.org/10.15376/biores.10.1.1543-1552
75. Xu, C., Ferdosian, F.: Lignin-based Phenol–Formaldehyde (LPF) resins/adhesives. Green Chem. Sustain. Technol. 91–109 (2017). https://doi.org/10.1007/978-3-662-54959-9_6
76. Yang, G., Gong, Z., Luo, X., et al.: Bonding wood with uncondensed lignins as adhesives. Nature **621**, 511–515 (2023). https://doi.org/10.1038/s41586-023-06507-5
77. Younesi-Kordkheili, H., Pizzi, A., Honarbakhsh-Raouf, A., Nemati, F.: The effect of soda bagasse lignin modified by ionic liquids on properties of the urea–formaldehyde resin as a wood adhesive. J. Adhes. **93**, 914–925 (2016). https://doi.org/10.1080/00218464.2016.1188284
78. Zhang, T., Xue, B., Yan, Q., et al.: New kinds of lignin/non-isocyanate polyurethane hybrid polymers: facile synthesis, smart properties and adhesive applications. Ind. Crops Prod. **199**, 116706 (2023). https://doi.org/10.1016/j.indcrop.2023.116706
79. Zhang, Y., Li, N., Chen, Z., et al.: Synthesis of high-water-resistance lignin-phenol resin adhesive with furfural as a crosslinking agent. Polymers **12**, 2805 (2020). https://doi.org/10.3390/polym12122805
80. Zhang, Y., Yuan, Z., Mahmood, N., Huang, S., Xu, C.: Sustainable bio-phenol- hydroxymethylfurfural resins using phenolated de-polymerized hydrolysis lignin and their application in bio-composites. Ind. Crops Prod. **79**, 84–90 (2016). https://doi.org/10.1016/j.indcrop.2015.10.048
81. Zhao, M., Jing, J., Zhu, Y., et al.: Preparation and performance of lignin–phenol–formaldehyde adhesives. Int. J. Adhes. Adhes. **64**, 163–167 (2016). https://doi.org/10.1016/j.ijadhadh.2015.10.010

Particleboard and Fiberboard Bonded with Green Binder

Nur Izzah Nabilah Haris, Syeed SaifulAzry Osman Al Edrus, Nur Haninah Harun, Arif Azizi Che Harun, Nurain Johar, and Mohd Hafizz Wondi

Abstract This chapter explores the innovative integration of green binders in lignocellulosic particleboards and fiberboards. The adoption of lignocellulosic materials not only mitigates wood resource scarcity but also converts agricultural waste into valuable resources. Green binders, including starch, tannins, citric acid, and lignin, emerge as eco-friendly alternatives, challenging the environmental impact of synthetic counterparts. The chapter also scrutinizes binderless technology, where the lignocellulosic material itself acts as the adhesive. Manufacturing processes, performance analyses, and the quest for a harmonious synthesis of sustainability and structural integrity characterize this exploration, offering valuable insights for researchers navigating the dynamic field of green binders in particleboard and fiberboard technology.

Keywords Green binder · Particleboard · Fiberboard · Binderless technology · Sustainable binder

N. I. N. Haris · S. S. Osman Al Edrus (✉)
Institute of Ecosystem Science Borneo, Universiti Putra Malaysia Bintulu Campus, Bintulu, Malaysia
e-mail: saifulazry@upm.edu.my

N. I. N. Haris · N. H. Harun
Department of Chemical and Environmental Engineering, Faculty of Engineering, Universiti Putra Malaysia, Serdang, Selangor, Malaysia

A. A. C. Harun · M. H. Wondi
Faculty of Plantation and Agrotechnology, Universiti Teknologi MARA Sarawak Branch Mukah Campus, Mukah, Malaysia

N. Johar
Faculty of Applied Sciences, Universiti Teknologi MARA Cawangan Sarawak Kampus Mukah, KM 7.5 Jalan Oya, Mukah 96400, Sarawak, Malaysia

S. S. Osman Al Edrus et al. (eds.), *Green Lignocellulosic-Based Panels*, Engineering Materials, https://doi.org/10.1007/978-981-96-3404-0_5

1 Introduction

The growing awareness of environmental sustainability, resource consumption, the global waste problem, and climate change has sparked a surge of interest in alternative and renewable resources. Lignocellulosic particleboard and fibreboard, strengthened by the incorporation of environmentally friendly binders, stands as an epitome of systematic collaboration between innovation and sustainability within the context of material science. The lignocellulose panel business today values items from renewable sources and biological origin, as well as binders used in their production line [17]. This synergistic approach has the capacity to not only reimagine traditional materials but also explore alternatives that uphold performance standards without compromising environmental integrity. This chapter embarks on a comprehensive exploration lignocellulosic particleboard and fibreboard, placing a spotlight on the concept of green binders.

Particleboard and fibreboard have garnered widespread popularity as construction materials such as floor underlayment, ceiling tiles, cupboards, cabinets, tabletops, and doors due to their versatility and cost-effectiveness [4]. Conventional boards are typically crafted from waste wood chips, sawmill shavings, or sawdust, combined with binders and compressed into composite materials [36]. However, the escalating demand for wood and land resulted in a scarcity of resources for wood-based particleboard and fibreboard manufacturing. Simultaneously, the thriving agricultural industry has generated abundant agricultural waste with no efficient disposal solution. The emergence of lignocellulosic particleboard and fibreboard is an ingenious solution, killing two birds with one stone. Not only does it alleviate the strain on limited wood resources, but it also transforms unwanted agricultural waste into a valuable resource. Although oil palm fibre and rice husk have conventionally taken the spotlight in particleboard production, continuous research is broadening the scope to assess the feasibility of utilizing various other agricultural wastes. Materials such as oil palm trunks, oil palm empty fruit bunch, coconut husk, hemp fibre, sunflower husk, sugar cane bagasse, coffee husk, and date palm fronds are presently undergoing scrutiny and development as potential raw materials within the board manufacturing industry [30].

The emergence of the adoption of green binders can be attributed to the growing awareness of environmental and health concerns in industrial processes. Commercial production of wood-based particleboard and fibreboard usually uses synthetic binders such as phenol-formaldehyde (PF), melamine-formaldehyde (MF), and urea-formaldehyde (UF) due to their cost effectiveness, strong binding properties, stability, and rapid curing capabilities [17]. However, these binders rely on non-renewable petroleum resources and contain harmful volatile organic compounds. Beyond depleting non-renewable resources, conventional binders release carcinogenic formaldehyde during manufacturing and usage of particleboard and fibreboard [30]. These issues have spurred the interest in exploring alternative binding agents that are environmentally sound and safe to humans. Starch, tannins, citric acid, latex, Arabic gum, gelatine, polylactic acid (PLA), polyhydroxy butyrate (PHB), are some examples of green binders that are used for lignocellulosic particleboard and fibreboards [17]. Numerous research has been conducted to explore the compatibility of lignocellulosic material with the binders and optimize the bonding properties by creating a suitable binder formulation. From another point of view, it is also possible to produce binderless particleboards and fibreboards, whereby the chemical component of the lignocellulosic itself will self-bond within the composite material. All these aspects of green binders will be further discussed in this chapter.

In essence, the integration of green binders with lignocellulosic particleboard and fibreboard represents a harmonious marriage of innovation and sustainability within the realm of material science. This chapter covers the multifaceted aspects of lignocellulosic composites, placing a spotlight on the pivotal concept of green binders. Furthermore, it delves into the manufacturing process and performance of particleboard and fibreboard bonded with green binders, shedding light on their application in practical settings. This chapter serves as a valuable reference for researchers and scientists seeking insight into green binders within the context of lignocellulosic particleboard and fibreboard technology.

2 Green Binders in Particleboard and Fiberboard

The worldwide bio-based or green binders market is expected to expand tremendously due to a significant increase in demand and acceptance by the lignocellulosic particleboard and fibreboard industries. The production of lignocellulosic particleboard generally involves the mixing of dried wood particles or plant-based materials with adhesive binders and pressed under heat and pressure forming a composite panel. The dried particles were sorted into fine and coarse sizes, for the skin and the core layer of composite panel, respectively. Engineered wood products are traditionally made with urea-formaldehyde (UF), phenol-formaldehyde (PF), melamine-urea formaldehyde (MUF), and melamine-formaldehyde (MF). However, the traditional adhesive binders used in commercial particleboard manufacturing often derived from petrochemical sources, pose environmental concerns due to their non-renewable nature and emissions during manufacturing [30]. Therefore, initiatives were taken to eliminate the usage of formaldehyde in producing particleboard and fibreboard. In response, the incorporation of green binders, derived from renewable and eco-friendly sources, has gained attraction.

Biobased adhesive binders, commonly derived from natural, non-mineral sources, serve as eco-friendly alternatives to synthetic adhesives. Among them, tannins, lignin, carbohydrates, unsaturated oils, proteins, dissolved wood, and self-adhesive wood welding stand at the forefront of the innovative development of green binders. Nonetheless, industrial technologies today predominantly rely on paraformaldehyde. Therefore, researchers have explored numerous experiments in an effort to cater to the demand for reducing formaldehyde emissions from bio-based adhesive binders, prioritizing user- and environmentally friendly attributes. There have been several works reported on the achievements of green binders in composite panel production, majorly in tannin, lignin, soy-based, carbohydrates, and citric acid.

2.1 *Tannin-Based Binders*

Tannin is the general term for a chemical that dissolves quickly in water and whose aqueous solution is very astringent, hence tanning leather, while, chemically, the term "tannin" refers to a broad class of organic compounds [16, 35]. Tannins are widely found in tree bark, wood, buds, leaves, stems, seeds, roots, fruits, and plant galls [15]. Extraction is the most common method for economically obtaining tannin from plant materials, such as wood, bark, leaf-like ethanol, methanol, acetone, ethyl acetate, and NaOH, using water alone or in combination with additional solvents [11, 18, 32]. In addition to classic tannin extraction using organic solvents and hot water, there are various alternative methods available, including ionic liquid extraction, pressurized hot water or subcritical water extraction, and microwave or ultrasonic extraction [7]. Condensed tannins are then copolymerized with other synthetic adhesives, resulting in lower formaldehyde emissions from synthetic resins in wood composites.

Amidst the effort to produce green binders, pure green binders were reported to have some constraints, in which additives are required to obtain a cured solid composite panel [31]. Therefore, combining and mixing the green binders with a hardener or another bio-based additive was considered to overcome the limitations. As reported by [27], a bio-hardener from African trees, *Vachellia nilotica*, and *Senegalia senegal*, has effectively strengthened a maritime pine tannin resin without any discernible aldehyde presence, resulting in entirely green binders for a composite panel. The gel times were slower compared to those achieved with paraformaldehyde but remained sufficiently fast to hold potential industrial significance.

Other researchers also reported on the formulation of green binders from the combination of mimosa tannin extract and lignin without the reinforcement of synthetic resins, (e.g., isocyanates or phenol-formaldehyde) to acquire the standard results [26]. The lignin used in this method was obtained from wheat straw pulping using formic acid/acetic acid and particleboards manufactured with these adhesive formulations have demonstrated compliance with the relevant interior standards. In addition to that, an attempt has been made with the combination of tannin extract and glucose, forming a range of non-toxic and non-volatile monomeric and oligomeric aldehydes [40]. These aldehydes reacted effectively with the tannin polyphenolic component, yielding satisfactory bonding outcomes for particleboards.

2.2 *Protein-Based Binders*

Protein-based binders, derived from soy-protein or soybean flour, have gained a reputation in composite panel production. These binders provide a renewable and biodegradable alternative to traditional petroleum-based adhesives. The manufacturing process involves converting soybean into a polymer that acts as a strong adhesive for binding wood particles or fibres in composite panels. Soy-based binders offer comparable performance to traditional binders, with the added benefit of being more environmentally friendly. Few works were reported in producing full bio-based adhesive binders from soy flour and protein sources.

The initial approach involves the interaction of a phenolic compound such as tannic acid with soy protein that produces a more viscous soy-based adhesive for a promising improved result in the panel board [14]. The other alternative method is formulating a soy-based adhesive through the reaction of soy protein isolate with maleic anhydride (MA) and hexamethylene diamine, introducing a completely novel green soy-based adhesive, soy protein isolate-based polyamides (SPIPA) [41]. In this work, SPIPA is then mixed again with MA as a cross-linking agent before the hot pressing step to produce a modified soy-based adhesive (SPIPA-MA), forming a dry and improved high-strength resistance composite panel that passes the panel standard of China (BG/T 9846–2015). The third notable technique reported by [13], focuses on obtaining entirely bio-based adhesives through the oxidation of a carbohydrate-protein mixture (soy flour). Additives such as dialdehydes (glyoxal and glutaraldehyde) were then added which demonstrated a higher wet shear strength and commendable bonding results for composite panel production.

2.3 *Carbohydrate-Based Binders*

Carbohydrate-based binders, sourced from polysaccharide-forming plants present another eco-friendly alternative. Three common groups used for green binders' production were derived from cellulose, starch, and gum. Due to its availability, low cost, and environmental friendliness, starch is a renewable green resource that is a significant biopolymer for modern industry (Chotikhun et al. 2018). Starch is a homopolysaccharide made up mostly of two polysaccharides (e.g. amylose and amylopectin). Amylose has a linear molecular morphology of -(1-4) glucose units and a low molecular weight, whereas amylopectin has a branching molecular morphology and a larger molecular weight. Figure 1 depicts several methods of starch modifications to make it suitable as an adhesive material.

Recently, bio-based binders produced by gelatinized corn starch modified with citric acid were reported by [23]. The cornstarch-based adhesive binder exhibits a significant result by passing the mechanical strength standard requirements by JIS A 5908. There is numerous research on green adhesive binders from furan resin, one kind of carbohydrate-based material produced from various furan compounds, commonly furfural alcohol and furfural synthesized from lignocellulose by dehydration of carbohydrates with acid catalyzation [2]. One of the outstanding works has been reported on the synthesis of green furan binders, namely, furfuryl alcohol-glyoxal (FAG) adhesive that were prepared by reacting furfuryl alcohol with aldehyde

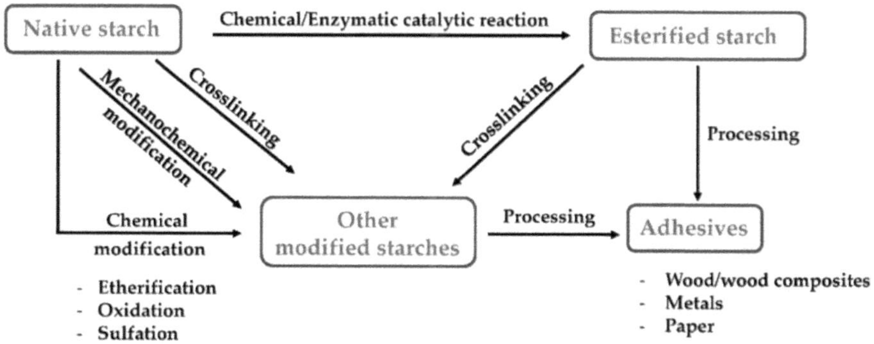

Fig. 1 Starch modification processes for adhesive applications. Reproduced with permission [39] [Creative Commons Attribution (CC BY) license]

(glyoxal) [42]. For panel preparation, furfuryl alcohol-aldehyde was mixed with p-Toluene sulfonic acid as a hardener for the bonding reaction. Based on this study, glyoxal successfully reacts with furfuryl alcohol to obtain a high-quality green binder. FAG adhesive binder has also passed the standard requirements for adhesive binders ($\geq$0.7 MPa).

In overview, while the adoption of green binders in composite panel production is steadily increasing, challenges such as cost, and scalability remain. Researchers and industry stakeholders are actively working to overcome these hurdles through innovations in manufacturing processes and the development of more cost-effective green binders. The ongoing research and advancements in green binder technologies draw attention to a promising future where composite panels can be produced with reduced environmental impact, contributing to a more sustainable and eco-friendly construction industry.

2.4 Lignin-Based Binders

Lignin, derived from grounded and sieved millings of diverse biomass sources such as forestry residues, agricultural remnants, and industrial by-products, serves as a promising contender for green binders in various applications [1, 24, 37]. Extracted through a meticulous delignification process, which encompasses alkaline pulping methods like Kraft and soda pulping or organosolv pulping, lignin undergoes a transformative journey. This process disrupts the ether and ester bonds within its larger structure, causing the resultant pieces to dissolve into a solution, forming a black liquid. Following pulping, the lignin is effectively separated by precipitating the black liquor [18] (Hussin et al. 2022). The unique physicochemical properties of lignin obtained through these eco-friendly processes position it as a versatile green binder, contributing to the sustainable development of materials in various industries.

To curb formaldehyde emissions, an experimental study focused on lightweight flax shive particleboards, using two binders: bio-based lignin (lignosulfonate) and partially bio-based epoxy resin (greenpoxy 56) [9]. With a target density of 500 kg/m^3, the particleboards underwent thermo-compression with 20 wt.% binder content and flax shives were directly used without pre-treatment to assess their raw material potential. Mechanical and flame tests revealed that lignosulfonate-based particleboards surpassed greenpoxy 56 in performance. However, greenpoxy 56 exhibited dimensional stability post-water immersion and commendable insulating properties. Both particleboards met EN 15,197 requirements for non-load-bearing flax boards in dry conditions (Type FB2), highlighting their competitiveness against urea-formaldehyde panels.

3 Binderless Particleboard and Fiberboard

The transition towards a circular and biobased economy demands alternatives to particleboard and fibreboard containing synthetic resins. The binderless board stands out as an interesting solution in the context of sustainable panels. Despite its indirect connection to the "green binders" discourse, the authors assert that the technology behind binderless boards significantly contributes to the creation of green panels, justifying its inclusion in this chapter. Binderless technology is the production of boards without the use of neither synthetic or green binders. Rather than relying on external binders to form a solid structure, the lignocellulosic raw material within the boards becomes the adhesive itself. This process involves heating and melting the components of the raw material such as lignin and cellulose, followed by compressing the melted components under high pressure. The self-bonding mechanism intrinsic to binderless technology stems from two distinct phenomena. First, there is adhesion between lignin-lignin and lignin-polysaccharide at elevated temperatures. Second, deformation occurs under high pressure at the "rubbery state" [25]. Eliminating external binders, however, comes with a trade-off. Binderless boards typically do not reach the standard of conventional bonded particleboards and fibreboards. Various studies have been done to investigate and improve the properties and strength of binderless boards.

Chen et al. [5] developed binderless particleboards from *Broussonetia papyrifera* trunk by hot pressing and examined the impact of particle size, pressing time and temperature, and board density on the properties. They found that the board met the Chinese national standard for heavy-duty particleboard under optimal manufacturing conditions. Their findings led them to propose that self-bonding comes from the degradation of cellulose and hemicellulose during high-temperature compression. Notably, the authors went the extra mile to demonstrate the bonding strength of the *Broussonetia papyrifera's* latex by bonding a poplar veneer with it. They proved that the latex had a significant bonding capability, indicating its contribution to the self-bonding of the binderless *Broussonetia papyrifera* particleboard.

In recent work, [33] fabricated binderless bamboo fibreboards, investigating the effect of different fibre sizes towards the void, water, and mechanical properties. The fibreboard was made into a customized shape with water as the carrier and hot-pressing. As water was used as a carrier during the fabrication process, effective water drainage from the matrix is deemed essential to produce a board with superior mechanical properties. The findings underscored the significance of incorporating larger bamboo fibres (<120 mesh) into the matrix to facilitate water release during hot-pressing. Conversely, smaller bamboo fibres (>120 mesh) were observed to cause bulging. The authors also proposed the formation mechanism of the binderless bamboo board, encompassing hydro, mechanical, and self-bonding effects (Fig. 2). The hydro effect comes from water acting as a medium during the fabrication. The presence of water during hot pressing reduced the softening temperature of cellulose, hemicellulose, and lignin, improving mechanical and chemical bonding between the fibres. The mechanical effect manifests when the matrix is compressed under high pressure, facilitating the diffusion of cellulose molecules and the formation of hydrogen bonds between hydroxyl groups. Lastly, self-bonding occurs when chemical bonds are formed between cellulose, hemicellulose, and lignin compounds.

In another study, [21] engineered binderless panels from oil palm trunks for insulation applications. They scrutinized the thermal and sound properties of the panels created with varying particle sizes and pressing durations. The results showed that larger particles improved the thermal and sound resistance, yet reduced the density, water resistance, and strength. Despite these trade-offs, the fabricated panels exhibit properties that could potentially be harnessed as thermal and sound insulation materials for buildings. Beyond these instances, several other examples showcase the creation of binderless boards from lignocellulosic materials such as rape straw [19], corncobs [29], giant reed [38], and sorghum [12]. References [25, 30, 37] provided comprehensive reviews on the binderless board technology from natural fibres including lignocellulosic materials. In essence, while the direct link to "green

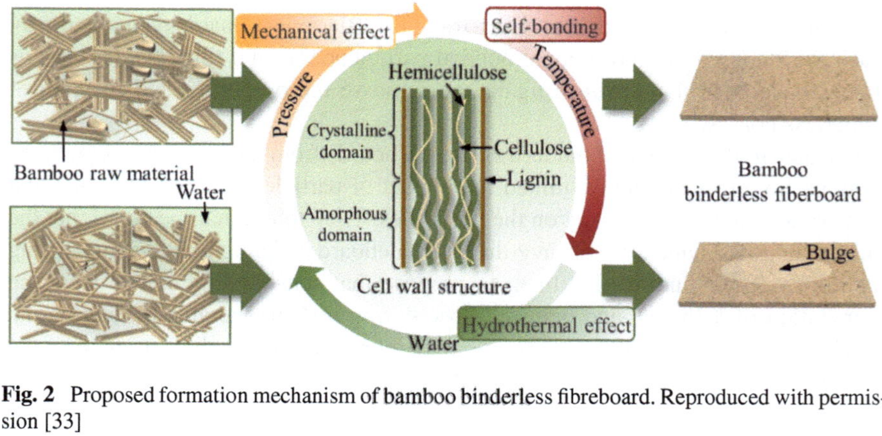

Fig. 2 Proposed formation mechanism of bamboo binderless fibreboard. Reproduced with permission [33]

binders" may not be immediately apparent, the inclusion of binderless board technology in this chapter signifies its pivotal role in advancing sustainability within the wooden panel sector.

4 Manufacturing Process of Particleboard and Fiberboard

Particleboard and fibreboard manufacturing involves the assembly of particles, fibres, or a combination of both, bound together by adhesive and subjected to specific press times and temperatures. The press, employing heat and pressure, activates the adhesive resin, resulting in a solid panel with commendable mechanical and physical properties, encompassing strength, stiffness, form, and dimensional stability [16]. Particleboards and fibreboards offer versatility in size and density, commonly available in sheet dimensions of 4' × 8' (1.22 m × 2.44 m) and 6' × 8' (1.83 m × 2.44 m), with thickness ranging from ¼ in. (6 mm) to 1 in. (25 mm). The density classifications include low-density (300 to 500 kg/m3 or 48 to 80 lb/ft^3), medium-density (500 to 650 kg/m^3 or 80 to 104 lb/ft^3), and high-density (650 to 700 kg/m^3 or 104 to 112 lb/ft^3) particleboards [34].

The manufacturing process comprises five pivotal steps: furnish preparation, resin application, mat formation, hot pressing, and finishing [3]. The furnish preparation stage, a cornerstone in particleboard and fibreboard manufacturing, involves refining raw materials into small fragments and meticulously drying them to achieve a moisture level ranging from 2 to 7% [3, 8]. This crucial step ensures the quality and uniformity of the particles used in the subsequent manufacturing process. Following furnish preparation, the resin application step becomes instrumental, where the choice of resin is dictated by the desired properties of the final product. The resin-to-wood ratio typically falls within the range of 6–9%, based on the resin's dry solids percentage and the particle's dry weight.

Once the particles are mechanically mixed with the adhesive system, the manufacturing process proceeds to mat formation, preparing the groundwork for the subsequent pressing stage. The material is then subjected to hot pressing at pressures ranging from 2 to 3 MPa and temperatures spanning from 140 to 220 °C, a critical phase in activating the adhesive resin and bonding the wood material into a solid panel. Upon completion of the press cycle, the panel undergoes a transition to a board cooler and is hot stacked, awaiting the final steps of cutting and sanding to achieve the desired finished panel sizes.

5 Performance of Particleboard and Fiberboard Bonded with Green Binder

Particleboard and fibreboard bonded with green binders offer a sustainable alternative with improved physical properties. These eco-friendly composite wood products boast improved dimensional stability, reduced moisture absorption, and optimal density, showcasing their potential for eco-conscious construction practices and favourable performance in various applications. In the study conducted by [10], the investigation focused on the formulation of a bio-based binder for particleboard production. Specifically, citric acid-based polyamide-epichlorohydrin (CA-PAE) was cross-linked with hydroxyl groups from keratin protein hydrolysate isolated from waste chicken feathers, with cellulose nanocrystals as a filler. The resulting adhesives demonstrated significant enhancements in the mechanical strength performance of the particleboard, meeting ANSI A208.1 standards, which include factors such as density, thickness, and strength.

The incorporation of green binders serves as a core point for clarifying the mechanical attributes and structural performance of particleboard and fibreboard, directing attention towards comprehending the influence of these eco-friendly adhesives on overall strength and durability. Research by [9] involved specific tests of bending, compression, and fire resistance tests towards their green bonded particleboard to gain a better understanding of how the boards perform in terms of strength and durability. The analytical outcomes reported that the particleboards formulated with bio-based lignin binders exhibit better mechanical and flame performances compared to their partially bio-based counterparts using greenpoxy 56. This enhanced performance can be attributed to the presence of lignosulfonate as a green binder, acting as a substantial internal adhesive agent that promotes superior cohesion within the particleboards. This delicate cohesion contributes to notable increases in both modulus and bending strengths. Interestingly, a similar cohesive effect leading to elevated modulus and bending strengths has been documented in bio-insulation materials based on bamboo powder and bio-binders, as reported by [28].

Understanding moisture absorption is crucial for predicting how boards respond to varying environmental conditions. Green-bonded particleboards and fibreboards exhibit a significant reduction in moisture absorption compared to their counterparts using traditional binders. Recent research, led by [22], involved the extraction of tannin from *Aningeria altissima* for green binder applications in the formulation of *Triumfetta cordifolia* needle-punched nonwovens fibreboards (TCF). Their work on developing new biocomposites bonded with extracted *Aningeria altissima* tannin showed high water sensitivity, surpassing the European standard EN 622–5, which specifies its use in dry conditions for construction and other applications. This underscores the potential of green-bonded boards to provide enhanced performance and durability in moisture-sensitive environments, particularly those associated with swelling or warping.

Additionally, outcomes from the same research indicated that the biocomposite bonded with hexamine-cured tannin resin exhibited optimal thermal stability, mechanical properties in the weft direction, high flexural strength, and flexural modulus in comparison to exudates from the *Vachellia nilotica* tree curing. These findings underscore the multifaceted properties of tannin-based biocomposites, demonstrating their potential for tailored applications in construction and related fields.

Essid et al. [9] conducted a comprehensive exploration into the heat resistance of particleboards utilizing the bio-based binder lignosulfonate (PLS) and a conventional, partially bio-based binder epoxy resin (PER). The study aimed to assess their performance under elevated temperatures. Results revealed that the thermal conductivity of PLS is comparable to PER, with mean values of 0.081 ± 0.001 and 0.074 ± 0.002 W/m.K, respectively. The slight difference was attributed to the pronounced porosity in PER, causing a reduction in thermal conductivity. This observation aligns with findings by Cravo et al. [6], elucidating the impact of high porosity on diminishing thermal conductivity in particleboards. Lower thermal conductivity means that the material is less effective at conducting heat. In applications such as building insulation or thermal barriers, this is advantageous as it helps maintain a more stable internal temperature by minimizing heat transfer.

The thermal conductivity of green-bonded particleboard is mainly influenced by manufacturing conditions, where variables such as pressing temperature, pressure, and duration play crucial roles. The type, concentration of green binders, and the board's density and moisture content further impact thermal conductivity. Liu et al. [20] emphasized that parameters affecting thermal conductivity include nominal density, manufacturing conditions (pressure and temperature), and raw material composition. Achieving optimal manufacturing conditions is utmost for enhancing particleboard's thermal performance. A meticulous balance in these factors ensures the effective integration of eco-friendly adhesives, yielding particleboards with enhanced thermal conductivity, suitable for applications requiring efficient insulation and heat transfer properties. Investigating and optimizing these manufacturing parameters contribute to the progression of sustainable building materials, aligning with contemporary environmental and performance standards.

The pursuit of developing particleboard and fibreboard bonded with green binders represents a progressive avenue for sustainable materials. However, this innovative approach may introduce variability in overall properties. Despite the promise inherent in transitioning towards eco-friendly adhesives, the challenge persists in striking an equilibrium between sustainability objectives and the preservation of structural integrity within this domain. The need to harmonize environmental considerations with the chemical, physical, mechanical, and thermal performance of these composite materials remains an ongoing challenge, necessitating further exploration and optimization to attain a cohesive synthesis of sustainability and structural robustness.

6 Conclusion

In conclusion, the exploration of green binders for lignocellulosic particleboard and fibreboard unveils a promising paradigm shift towards sustainable and innovative materials. The emergence of green binders, prompted by environmental and health awareness, challenges the conventional use of synthetic binders that rely on non-renewable resources and release harmful compounds. The comprehensive examination of tannin, protein, carbohydrate, and lignin-based binders elucidates the versatility and potential of these green alternatives. Challenges such as cost and scalability remain, but the concerted efforts of researchers and industry stakeholders signify a trajectory towards overcoming these hurdles. The manufacturing process intricately involves furnish preparation, resin application, mat formation, hot pressing, and finishing. Understanding these stages is crucial in ensuring the quality and uniformity of the final product. The performance of particleboard and fibreboard bonded with green binders showcases improved dimensional stability, reduced moisture absorption, and optimal density, aligning with the principles of eco-conscious construction practices. While the path towards green binders introduces variability in material properties, the challenge lies in harmonizing environmental considerations with the chemical, physical, mechanical, and thermal performance of these composite materials. Further investigations are crucial to strike a balance between sustainability objectives and structural robustness. In essence, this chapter serves as a valuable reference for researchers and scientists delving into the dynamic realm of lignocellulosic particleboard and fibreboard technology, emphasizing the transformative potential of green binders in reshaping the landscape of sustainable materials.

References

1. Armah, E.K., Chetty, M., Rathilal, S., Asante-Sackey, D., Tetteh, E.K.: Lignin: value addition is key to profitable biomass biorefinery (Chap. 11, Sahay SBT-H of B (ed.)). Academic, pp. 233–247 (2022)
2. Ashokkumar, V., Venkatkarthick, R., Jayashree, S., Chuetor, S., Dharmaraj, S., Kumar, G., Chen, W.-H., Ngamcharussrivichai, C.: Recent advances in lignocellulosic biomass for biofuels and value-added bioproducts - a critical review. Bioresour. Technol. **344**, 126195 (2022)
3. Carvalho, L., de Magalhães, F., Ferra, J.M.: Formaldehyde emissions from wood-based panels: testing methods and industrial perspectives (2012)
4. Chaturvedi, S., Verma, A., Sethi, S.K., Mavinkere Rangappa, S., Siengchin, S.: Stalk fibers (rice, wheat, barley, etc.) composites and applications (Chap. 15). In: Mavinkere Rangappa, S., Parameswaranpillai, J., Siengchin, S., Ozbakkaloglu, T., Wang their Composites, and Applications HBT-PF (eds.) The Textile Institute Book Series. Woodhead Publishing, pp. 347–362 (2022)
5. Chen, M., Zheng, S., Wu, J., Xu, J.: Study on preparation of high-performance binderless board from Broussonetia papyrifera. J. Wood Sci. **69**, 17 (2023)
6. Cravo, J.C.M., de Lucca, S.D., Mármol, G., Schmidt, G.M., de Carvalho Balieiro, J.C., Fiorelli, J.: Effect of density and resin on the mechanical, physical and thermal performance of particleboards based on cement packaging. Constr. Build. Mater. **151**, 414–421 (2017)

7. de Hoyos-Martínez, P.L., Merle, J., Labidi, J., Charrier – El Bouhtoury, F.: Tannins extraction: a key point for their valorization and cleaner production. J. Clean. Prod. **206**, 1138–1155 (2019)
8. Dunky, M.: Adhesives in the wood industry. In: Handbook of Adhesive Technology. CRC Press, Florida, p. 64 (2017)
9. Essid, S., Hegde, V.J., Mahieu, A., Bizet, L., Leblanc, N., Saouab, A.: Comparison of the properties of flax shives based particleboards prepared using binders of bio-based lignin and partially bio-based epoxy resin. Int. J. Adhes. Adhes. **109**, 102915 (2021)
10. Fagbemi, O.D., Sithole, B.: Evaluation of waste chicken feather protein hydrolysate as a bio-based binder for particleboard production. Curr. Res. Green Sustain. Chem. **4**, 100168 (2021)
11. Farahmandfar, R., Asnaashari, M., Bakhshandeh, T.: Influence of ultrasound-assist and classical extractions on total phenolic, tannin, flavonoids, tocopherol and antioxidant characteristics of Teucrium polium aerial parts. J. Food Meas. Charact. **13**, 1357–1363 (2019)
12. Ferrandez-Garcia, A., Ferrandez-Garcia, M.T., Garcia-Ortuño, T., Ferrandez-Villena, M.: Influence of the density in binderless particleboards made from sorghum. Agronomy **12**, 1387 (2022)
13. Frihart, C.R., Lorenz, L.F.: Specific oxidants improve the wood bonding strength of soy and other plant flours. J. Polym. Sci. Part A Polym. Chem. **57**, 1017–1023 (2019)
14. Ghahri, S., Mohebby, B., Pizzi, A., Mirshokraie, A., Mansouri, H.R.: Improving water resistance of soy-based adhesive by vegetable tannin. J. Polym. Environ. **26**, 1881–1890 (2018)
15. Giovando, S., Koch, G., Romagnoli, M., Paul, D., Vinciguerra, V., Tamantini, S., Marini, F., Zikeli, F., Scarascia Mugnozza, G.: Spectro-topochemical investigation of the location of polyphenolic extractives (tannins) in chestnut wood structure and ultrastructure. Ind. Crops Prod. **141**, 111767 (2019)
16. Gonçalves, D., Bordado, J.M., Marques, A.C., dos Santos, R.: Non-formaldehyde, bio-based adhesives for use in wood-based panel manufacturing industry-a review. Polymers (Basel) **13**, 4086 (2021)
17. Gumowska, A., Robles, E., Kowaluk, G.: Evaluation of functional features of lignocellulosic particle composites containing biopolymer binders. Materials (Basel) **14**, 7718 (2021)
18. Hussin, M.H., Rahim, A.A., Mohamad Ibrahim, M.N., Brosse, N.: Physicochemical characterization of alkaline and ethanol organosolv lignins from oil palm (Elaeis guineensis) fronds as phenol substitutes for green material applications. Ind. Crops Prod. **49**, 23–32 (2013)
19. Jerman, M., Böhm, M., Dušek, J., Černý, R.: Effect of steaming temperature on microstructure and mechanical, hygric, and thermal properties of binderless rape straw fiberboards. Build. Environ. **223**, 109474 (2022)
20. Liu, K., Takagi, H., Osugi, R., Yang, Z.: Effect of lumen size on the effective transverse thermal conductivity of unidirectional natural fiber composites. Compos. Sci. Technol. **72**, 633–639 (2012)
21. Mawardi, I., Aprilia, S., Faisal, M., Ikramullah, R.S.: An investigation of thermal conductivity and sound absorption from binderless panels made of oil palm wood as bio-insulation materials. Results Eng. **13**, 100319 (2022)
22. Mewoli, A.E., Segovia, C., Njom, A.E., Ebanda, F.B., Eyinga Biwôlé, J.J., Xinyi, C., Ateba, A., Girods, P., Pizzi, A., Brosse, N.: Characterization of tannin extracted from Aningeria altissima bark and formulation of bioresins for the manufacture of Triumfetta cordifolia needle-punched nonwovens fiberboards: novel green composite panels for sustainability. Ind. Crops Prod. **206**, 117734 (2023)
23. Mohamad Amini, M.A., Hashim, R., Sulaiman, N.S., Mohamed, M., Sulaiman, O.: Citric acid-modified starch as an environmentally friendly binder for wood composite making. BioResources **15**, 4234–4248 (2020)
24. Mujtaba, M., Fernandes Fraceto, L., Fazeli, M., Mukherjee, S., Savassa, S.M., Araujo de Medeiros, G., do Espí, A., Mancini, S.D., Lipponen, J., Vilaplana, F.: Lignocellulosic biomass from agricultural waste to the circular economy: a review with focus on biofuels, biocomposites and bioplastics. J. Clean. Prod. **402**, 136815 (2023)
25. Nasir, M., Khali, D.P., Jawaid, M., Tahir, P.M., Siakeng, R., Asim, M., Khan, T.A.: Recent development in binderless fiber-board fabrication from agricultural residues: a review. Constr. Build. Mater. **211**, 502–516 (2019)

26. Navarrete, P., Mansouri, H.R., Pizzi, A., Tapin-Lingua, S., Benjelloun-Mlayah, B., Pasch, H., Rigolet, S.: Wood panel adhesives from low molecular mass lignin and tannin without synthetic resins. J. Adhes. Sci. Technol. **24**, 1597–1610 (2010)
27. Ndiwe, B., Pizzi, A., Tibi, B., Danwe, R., Konai, N., Amirou, S.: African tree bark exudate extracts as biohardeners of fully biosourced thermoset tannin adhesives for wood panels. Ind. Crops Prod. **132**, 253–268 (2019)
28. Nguyen, D.M., Grillet, A.-C., Bui, Q.-B., Diep, T.M.H., Woloszyn, M.: Building bio-insulation materials based on bamboo powder and bio-binders. Constr. Build. Mater. **186**, 686–698 (2018)
29. Orisaleye, J.I., Abiodun, Y.O., Ogundare, A.A., Adefuye, O.A., Ojolo, S.J., Jekayinfa, S.O.: Effects of manufacturing parameters on the properties of binderless boards produced from corncobs. Trans. Indian Natl. Acad. Eng. **7**, 1311–1325 (2022)
30. Owodunni, A.A., Lamaming, J., Hashim, R., Taiwo, O.F.A., Hussin, M.H., Mohamad Kassim, M.H., Bustami, Y., Sulaiman, O., Amini, M.H.M., Hiziroglu, S.: Adhesive application on particleboard from natural fibers: a review. Polym. Compos. **41**, 4448–4460 (2020)
31. Pizzi, A., Papadopoulos, A.N., Policardi, F.: Wood Composites and Their Polymer Binders. Polymers (Basel) **12**, 1115 (2020)
32. Sepperer, T., Hernandez-Ramos, F., Labidi, J., Oostingh, G.J., Bogner, B., Petutschnigg, A., Tondi, G.: Purification of industrial tannin extract through simple solid-liquid extractions. Ind. Crops Prod. **139**, 111502 (2019)
33. Shi, J., Zhong, T., Xu, X., Wu, J., Zou, Y., Fei, B., Xia, C., Chen, H.: Eco-friendly and special-shaped bamboo binderless fiberboards fabricated by self-bonding technology: effect of bamboo fibers with different sizes. Ind. Crops Prod. **194**, 116300 (2023)
34. Suana Sugahara, E., Mello, A., da Silva, S., Soler Cunha Buzo, A., Inácio de Campos, C., Martines Morales, E., Santos Ferreira, B., dos Anjos, A.M., Rocco Lahr, F.A., Christoforo, A.L.: High-density particleboard made from agro-industrial waste and different adhesives. BioResources **14**, 5162–5170 (2019)
35. Tahir, P.M., Halip, J.A., Hua Lee, S.: Tannin-based bioresin as adhesives (Chap. 7, Ariffin, H., Sapuan, S.M., Hassan MABT-L for FB (eds.)). Elsevier, pp. 109–133 (2019)
36. Vale, B.: 3 - Building materials. In: Petrović, E.K., Vale, B., Zari Ecological and Sustainable Built Environment MPBT-M for a H (eds.) Woodhead Publishing Series in Civil and Structural Engineering. Woodhead Publishing, pp. 67–112 (2017)
37. Vitrone, F., Ramos, D., Ferrando, F., Salvadó, J.: Binderless fiberboards for sustainable construction. Materials, production methods and applications. J. Build. Eng. **44**, 102625 (2021)
38. Vitrone, F., Ramos, D., Ferrando, F., Salvadó, J.: The influence of thickness and final heat treatment on mechanical properties and dimensional stability of binderless fiberboards from steam exploded Arundo donax L. Mater. Today Proc. **58**, 999–1004 (2022)
39. Watcharakitti, J., Win, E.E., Nimnuan, J., Smith, S.M.: Modified starch-based adhesives: a review. Polymers (Basel) **14**, 2023 (2022)
40. Xi, X., Pizzi, A., Frihart, C.R., Lorenz, L., Gerardin, C.: Tannin plywood bioadhesives with non-volatile aldehydes generation by specific oxidation of mono- and disaccharides. Int. J. Adhes. Adhes. **98**, 102499 (2020)
41. Xi, X., Pizzi, A., Gerardin, C., Chen, X., Amirou, S.: Soy protein isolate-based polyamides as wood adhesives. Wood Sci. Technol. **54**, 89–102 (2020)
42. Xi, X., Wu, Z., Pizzi, A., Gerardin, C., Lei, H., Du, G.: Furfuryl alcohol-aldehyde plywood adhesive resins. J. Adhes. **96**, 814–838 (2020)

Performance of Wood Polymer Composite Using Eco-Friendly Binders

Siti Norasmah Surip, Falah Abu, Mohd. Nazarudin Zakaria, Siti Rafedah Abdul Karim, and Azrena Abdul Karim

Abstract Wood-plastic composites (WPCs) are composite materials made from a combination of wood fibers or particles and thermoplastics, such as polyethylene, polypropylene, and polyvinyl chloride. The combination of wood and plastic results in a material that has the look and feel of wood, but with added durability, water resistance, and reduced maintenance requirements. The production of WPCs involves mixing the wood particles or fibers with binders, and then processing the mixture into the desired shape, such as boards, profiles, and tiles. Binders play an important role in the performance of WPC materials. They provide the necessary adhesive strength to hold the composite together and can also help improve the mechanical properties of the material, such as its strength and stiffness. The type and amount of binder used can also impact the final properties of the WPC, such as its water resistance and resistance to decay. Eco-friendly binders are often used in wood-plastic composites (WPCs) as an alternative to traditional petroleum-based binders. Some of the most commonly used eco-friendly binders in WPCs include starch and lignin. Overall, eco-friendly binders offer a more sustainable and environmentally friendly option for binding wood particles in WPCs. They can help reduce the dependence on petroleum-based binders and improve the overall sustainability of WPCs.

Keywords Binders · Eco-friendly · Composites · Wood · Fibers

1 Introduction

Wood-plastic composites (WPC) have gained popularity in recent years as a sustainable alternative to traditional building materials. However, the binders used to hold the WPC together can have negative environmental impacts. As consumers become increasingly conscious of their environmental footprint, there is a growing demand

S. N. Surip (✉) · F. Abu · M. N. Zakaria · S. R. A. Karim · A. A. Karim
Faculty of Applied Sciences, Eco-Technology Program, UniversitiTeknologi MARA, Shah Alam, Selangor, Malaysia
e-mail: snorasmah@uitm.edu.my

S. S. Osman Al Edrus et al. (eds.), *Green Lignocellulosic-Based Panels*, Engineering Materials, https://doi.org/10.1007/978-981-96-3404-0_6

for eco-friendly binders for WPC. In this chapter, we will explore some of the available options for eco-friendly binders and their potential benefits. One promising option for eco-friendly binders in WPC is bio-based polymers. These polymers are made from renewable resources such as starch, cellulose, and lignin as shown in Fig. 1. They can be used as a replacement for petroleum-based binders, which are non-renewable and can have negative environmental impacts. Bio-based polymers have been found to be effective binders for WPC, with comparable mechanical properties to traditional binders [14]. Another option for eco-friendly binders in WPC is the use of recycled materials. Binders made from recycled plastics or other materials can help reduce waste and prevent these materials from ending up in landfills or the ocean. Some companies have already started using recycled binders in their WPC products, such as Trex Company, a manufacturer of eco-friendly decking [32].

A third option for eco-friendly binders in WPC is the use of natural adhesives. Natural adhesives, such as those made from soy or corn, are biodegradable and can have lower environmental impacts than traditional petroleum-based adhesives. However, they may have lower mechanical properties than other binders, and further research is needed to determine their effectiveness in WPC [1]. These eco-friendly binders have the potential to reduce the environmental impacts of WPC and meet the growing demand for sustainable building materials. However, it is important to consider the entire life cycle of WPC products and their environmental impacts beyond just the binder used. Overall, eco-friendly binders are a sustainable alternative to traditional petroleum-based binders. They are made from renewable or recycled materials and have a lower environmental impact. Using eco-friendly binders can help

Fig. 1 Sources of eco-friendly binders from renewable materials

reduce the carbon footprint of various industries and contribute to a more sustainable future.

2 Starch-Based Binders of Wood-Plastic Composites

The use of wood-plastic composites (WPCs) in various applications has gained much attention in recent years due to their unique properties and eco-friendliness. However, the binding agents used in the production of WPCs can have a significant impact on their properties and performance. One of the most commonly used binders in WPCs is starch-based binders. The morphology of starch-based binders plays a crucial role in determining the properties of WPCs. The overall structure of the starch-based binder is determined by various factors such as the plant source, geographic growth location, soil type and condition, growing climate conditions, and its amylose/amylopectin ratio, which in turn affect its morphology [22].

Starch is a cost-effective, abundant, renewable, and biodegradable polymer that fulfills the criteria for an ideal material. It is mainly extracted from the seeds, stalks, and roots of food staples like rice, corn, wheat, tapioca, and potato, and is the second most common natural polymer, following cellulose [10]. Starch granules are the primary form in which carbohydrates are stored in seeds and underground plant parts, but they are also present in smaller amounts in leaves and bark. When leaves are exposed to sunlight, transitional starch granules can be seen in the chloroplast stroma and cytoplasm, and they are transported to other organs for storage during periods of darkness. The physical attributes of starch granules, including their size, shape, and amylose to amylopectin ratio, vary among different biological sources. Glycans are a type of carbohydrate that includes polysaccharides like starch. Starch is composed of D-glucose, a six-carbon sugar, which serves as its "building block". D-glucose, a monosaccharide, can be represented in graphical form either as a straight chain or a ring. Starches differ in composition and structure due to genetic, environmental, and nutritional factors. Amylose, lipids, phosphorylated residues, and long lateral chain amylopectin interact to avoid water uptake, while short lateral chain amylopectin allows hydration and produces gels that can retrograde. Smaller starch granules have a larger surface area for faster hydration, but surface pores and channels increase water uptake, leading to increased swelling, viscosity, and gelatinization. The chemical compounds and interactions on the starch granule also affect its thermal properties [4].

There have been relatively few investigations carried out on the subject of the physical and mechanical properties of starch-based binders used in wood-plastic composites. Serra-Parareda et al. [27] in their study tried to improve the PLA properties by incorporating PTA (thermoplastic starch-based polymer) reinforced with bleached kraft hardwood fibers (BKHF) to improve their mechanical performance. It was found that for the mixture containing 20% PTA and reinforced with 30% BKHF, the tensile strength increased by 35.4% to 51.32 MPa, Young's modulus increased by 112.3–5.54 GPa, and the strain increased by 23.5–4%. Meanwhile, for thermal, morphological, and macro-mechanical properties it was found that the tensile properties and the interphase of the two components were fairly good in comparison with the literature.

In a recent study, Bartolome et al. [2] used enzymatically polymerized lignosulfonates and carboxylic acids as additives to enhance the wet resistance and bonding capabilities of starch-based adhesives. Using laccase, lignosulfonates underwent polymerization, resulting in a molecular weight of 750 kDa. By including small amounts (up to 1% of the starch weight) of 1,2,3,4-butanetetracarboxylic acid (BTCA), the adhesive properties were enhanced, but the incorporation of higher amounts of BTCA reduced the measured properties due to a significant increase in viscosity. The adhesive that contained 8% enzymatically polymerized lignin and 1% BTCA exhibited remarkable enhancements in wet-resistance (from 22 to 60 min) and bonding times (from 30 to 20 s). However, the addition of citric acid (CA) had an adverse effect on the adhesive properties, particularly in the presence of lignosulfonate. A similar trend was also observed by [16] in their study.

In another study, [21] investigated the effect of various wt/wt concentrations of starch (tapioca) and polypropylene to the rheological and mechanical properties of WPCs. They discovered that increasing starch concentrations reduced the melt flow index linearly up to 30 wt% starch to wt% plastic, after which no flow was observed. The mechanical properties were found to decrease with starch loadings above 30%, while some mechanical properties improved with lower starch loadings below 30%. Furthermore, a higher starch loading above 30% has an adverse effect on the melt flow index and the Izod impact strength, as there was no flow due to a lower softening temperature.

Another study by [7] examined the effect of two types of natural binders and its content on the mechanical and thermal properties of Typha Leaf powder panels. According to their findings, the typha-starch board with a 50% binder exhibited a favorable compressive strength of 1.05 MPa, which is 1.36 times stronger than the typha-gum arabic board. Additionally, the optimal thermal conductivity was observed with the typha-starch board, containing 33.33% binder, which had a thermal conductivity of 0.051 W m^{-1} K^{-1}. The thermal conductivity values are the same as or lower than those of many natural insulators.

3 Properties and Applications of Lignin-Based Bonded Wood Polymer Composite

At the forefront of new developments of bio-based binders, including those materials of natural, non-mineral origin, research and utilization of renewable bio-based binders in WPC manufacturing are increasingly reported. These include carbohydrates/starch, hydrolyzable tannins, lignin, unsaturated oils, soy protein, proteins and protein hydrolysates, and dissolved wood [23, 24]. Research and development of lignin-based adhesive system were reviewed by many researchers [12, 20, 24, 29,

34, 36]. At a commercial level, the use of pure lignin is not successful due to, among others, too long a pressing time and high corrosiveness for the equipment, hence they did not meet with commercial standards.

Formylated lignin addition (up to 20%) has been used extensively in the plywood industry in Canada. More recently, an LPF resin has been commercialized for plywood containing up to 50% lignin and improvements on this by using ionic liquid-treated lignin to prepare LPF, and LP-glyoxal resins have also been reported with encouraging results [34]. Soy protein, gelatine, zein protein, pectin, and Salix lignin provided wet tear strength equivalent to that of the vinyl acetate binder [9]. In the same report, gelatine showed a slight increase in strength when horseradish peroxidase was added.

Formaldehyde-based adhesives are being reduced by stricter government regulations, consumer awareness, and the resulting drive of the industry favoring more environmentally friendly materials, at the same time reducing the dependency on fossil resources. UF-starch, esterified starch, and oxidized starch are among green binders reported. Lignin has a lower cost and is more abundant and more easily available than all other bio-based polymers. Hence, new formulations of wood adhesive based on lignin are increasing in number such as lignin-phenol-formaldehyde (LPF), lignin-phenol-glyoxal (LPG), lignin-urea-formaldehyde (LUF), lignin-urea-glyoxal (LUG), and others [36].

3.1 Application of Lignin in Wood Polymer Composites

Majority of the world's lignin supply has come from pulping waste or Kraft-lignin. Kraft-lignin, a by-product, originated from the black liquor in the pulp and paper industry as a source of raw material for making WPC in many parts of the world. It is worth noting that lignin is readily separated from the other polysaccharides in the pulping process, and they are readily available for production. Conventionally, utilization of lignin by-products is limited as fuel sources in the pulp and paper mills. Other lignin sources are sorghum, organosolv-lignin, Eucalyptus, and bamboo [24].

Lignin is a highly complex heterogeneous polymer, found naturally in plants with woody biomass. Lignin is made up of C_6C_3 phenolic units that are linked by cross-linking, containing phenolic compounds including aliphatic and aromatic hydroxyl, methoxyl, carboxyl, or carbonyl. A drawback related to its complexity is due to its changeable macromolecular structure. One way to deal with this problem is to convert unordered raw lignin into uniform nanoparticles, which could be promising candidates for various applications. Several researchers inferred the potentiality of the use of nanolignin particles as stabilizers, reinforcing additives, antioxidants, and nanocarriers. This is due to their advantages of being biocompatible, non-toxic, thermally stable, and environmentally durable [37].

Lignin phenolic compounds cross-link in a variety of ways to give among others the thermal stability, biodegradable resistance, and increased UV absorbance. In the conventional WPC, bonding of biopolymer and matrix relies not only on the effect of synthetic resins but also on the intrinsic auto-adhesive or self-bonding properties of wood materials [12]. Since synthetic resin is toxic, a more environmentally friendly approach to boost up the inherent capacity of wood elements to bind among themselves were reported [12, 24, 37].

Approaches in increasing the self-bonding capacity of lignin compounds were reported, including the oxidative modification, oxidative polymerization, and enzymatic cross-linking [12, 24, 37]. It is noted that enzymatic cross-linking mimics the natural process of lignification in living plants and trees [12]. The specific use of phenol-oxidizing enzymes including laccases and peroxidases to promote the polymerization and cross-linking of lignin was reported. Modifications and cross-linking to improve self-bonding of lignin structure were also reported by [33].

In the manufacture of WPC, lignin is added to UF and PF resin; therefore the use of lignin as a substitute for phenol in PF resins is growing [35]. Previous studies have shown that lignin can cross-link with formaldehyde and reduce formaldehyde emission from the panels made from LUF and LPF resins, while not significantly affecting the bonding strength of the bonded panels. It was also shown that the addition of 20% lignin to a UF resin can reduce its formaldehyde emission from 5.4 mg/100 g to 3 mg/100 g in particleboard panels [36]. However, the main obstacle to increasing lignin application to phenolic adhesives is due to its slower reaction than phenol with formaldehyde. For this reason, it is proven that lignin must be modified before any use of it in the PF resin for wood adhesive. Glyoxalated-lignin, phenolated-lignin, ionic-treated lignin, and Maleic anhydride-modified lignin are modified lignin reported [35].

Binderless fiberboard manufacturing process applying the oxidative modification of lignin to enhance self-binding capacity was reported [12]. Several studies have reported the use of lignin-based adhesives in many applications, notably in wood polymer composites. It has also been reported that lignin is used as reinforcement filler, adhesion enhancer/modifier, additives, dispersants, coupling agents, antioxidants, antimicrobial agents, flame retardants, and carbon precursors [12, 20, 24, 29, 34, 36].

In terms of properties, WPCs suffer from the problem of low compatibility between the hydrophobic polymer matrix and the hydrophilic wood filler. The improvement of the adhesion affinity between binders and wood fibers in WPCs can be achieved by the modification of the fiber surface and/or the application of specific functional additives such as compatibilizers, coupling agents, and adhesion enhancers [29]. The functional additives can improve the properties of the filler particles, thus preventing their agglomeration and improving the adhesion interaction between a wood filler and a polymer matrix at the interface. They are normally applied in a small amount of 3–5 wt% by dry mass of the composite [29]. In the last decades, silanes, isocyanates, maleic anhydride-grafted polypropylene (PP-g-MAH), and polyethylene (PE-g-MAH), polyethylenimine (PEI) and maleated polyolefins have been applied as interface adhesion modifiers for the formulation of WPCs [29, 37]. Unfortunately, the most used functional additives for improving interfacial adhesion at the filler/polymer interface are synthetic, high-priced, and not biodegradable [29].

A small percentage of lignin as dispersants and fillers has the potential to enhance various properties including the mechanical, thermal, thermo-chemical, anticorrosive, flame retardant, and other properties of wood polymer composite. The presence of polar and non-polar sites in the lignin structure, known as hydrophilic and hydrophobic sites, indicates that it can be used as the coupling agent in WPCs. As compared to the synthetic coupling agent such as silane and maleic anhydride polypropylene (MAPP), the effectiveness of lignin as coupling agent is lower; thus modification is needed. Modification of lignin into nanolignin was reported [36]. A higher strength and dimensional stability were obtained with nanolignin-containing composites as compared to the normal lignin-containing composites. It was also stated that nanolignin exhibited comparable properties with the control samples prepared using MAPP. Summary of the function of pure lignin and modified lignin in WPC is presented in Table 1.

3.2 Physico-Mechanical Properties and Morphology Characteristics of Lignin-Bonded WPCs

In general, the literature indicated that addition of pure or modified lignin as binder or additive in the composite mixture results in good physical and mechanical properties of WPC. It is to be noted that the main problem associated with lignin-based adhesive is that lignin is complex and heterogeneous. Hence, different lignin with different inherent physico-chemical properties have a different effect on the properties of the resins prepared [36]. For example, heterogeneity of lignin from pulping by-products mainly depends on the pulping process and method used to separate the lignin from the cooking liquor such as Sulfite, Organosolve, Soda, and Kraft. Some of the physico-mechanical and morphological properties of WPC prepared with addition of lignin-based binder or additives are presented here.

Physico-mechanical properties

Comparisons of the physical and mechanical properties of particleboard made from LPF and LUP resin were reported [36]. Lignin is obtained from the Kraft and Soda process. LPF have higher mechanical strength and dimensional stability compared to those made from LUF resins at constant level of lignin. The particleboards bonded with Soda-based resin yielded lower water absorption and formaldehyde emission as well as higher flexural strength and internal bonding strength when compared to those bonded with the Kraft-based resin. The mechanical properties of the particleboards bonded with LPG resins are significantly improved by increasing percentages of lignin, while they were decreased for LUF resin.

A study by [20] is referred to herein. Compared with unmodified lignin (O-lignin), functionalized lignin (F-lignin) is more effective to improve the thermal stability and flame retardancy of WPC because of presence of the flame retardancy elements (P and N) and the catalytic effect of Cu^{2+} on the char-formation. This study by Liu

Table 1 Function of lignin in WPC

Lignin added to other substances	Type of WPC/Pulp and paper	Source
Lignin/ammonium polyphosphate-modified wood flour as flame retardant	High-density polyethylene composites	Wei et al. [34]
i. Nanolignin as coupling agent ii. 5 wt% of lignin nanoparticles to improve dimensional stability and higher mechanical strength	Polypropylene-wood flour composites	Younesi-Kordkheili and Pizzi [36]
Lignin-Modified-LPF (Lignin Phenol Formaldehyde) Resin as Wood Adhesive	Particleboard	Younesi-Kordkheili and Pizzi [35]
Horseradish peroxidase and laccase-modified Kraft lignin (enzymatic modification of lignin to promote cross-linking for better adhesion)	Paper	Flory et al. [9]
i. Nitrogen and phosphorus-grafted soda lignin ii. Metal ions-grafted soda lignin Lignin-based flame retardants for enhancing the thermal and fire retardancy	PP-wood composite	Liu et al. [20]
Ammonium polyphosphate-mixed lignin as flame retardant to prevent the combustion of wood flour/high density polyethylene (WF/HDPE) composites	HDPE composite	Wei et al. [34]
Lignin-synthesized PF and UF (LPF, LUF) as physico-mechanical property enhancer	Particleboard	Younesi-Kordkheili and Pizzi [37]
i. Lignin-containing polyelectrolyte complexes (LPEC) nanoparticles ii. The treatment with the LPEC nanoparticles led to an increase in the fixed nitrogen content in aspen saw-dust microparticles, thus decreasing the amount of adsorbed water and increasing the contact angle	PP-aspen saw-dust composite	Shulga et al. [31]
Laccase enzyme-modified Kraft lignin as cross linking enhancer	Medium density fiberboard	Gouveia et al. [12]

et al. suggests a novel green strategy for improving flame retardancy performance of WPC and promoting the utilization of industrial lignin. Wei et al. [34] studied Lignin/Ammonium Polyphosphate (APP)/Wood Flour/HDPE (L/APP/WF/HDPE) composite. In this study, lignin was combined with APP in a 1:1 ratio and added to the WF/HDPE composite as flame retardant. The limit oxygen index of the L/APP/WF/HDPE composites rose with increasing L/APP amount. The carbon layer formed by combustion became denser due to the addition of L/APP, and oxygen could not easily enter the core layer. Thus, the amount of oxygen required to support combustion decreased. When the total amount of L/APP reached 20%, the composite had a LOI of 26.6, but when the L/APP content was increased to 50%, the LOI increased to 31.1, However, when the amount of flame retardant was increased to 50%, the tensile strength, bending strength, and impact strength decreased by 52.2, 42.5, and 48.3% respectively, compared with the pure WF/HDPE composite.

Younesi-Kordkheili and Pizzi [37] revealed that adding nanolignin coupling agent into PP composite improves both the bending modulus and bending strength. Results from the micrograph showed that there is an excellent dispersive characteristic of nanolignin in the PP composite. Progressively increasing the lignin nanoparticle proportion from 0 wt% to 5 wt% resulted in better dimensional stability and higher mechanical strength. The panels prepared with Soda-based lignin had higher flexural strength and dimensional stability compared to those made from the resins containing Kraft lignin.

Morphology characteristics

Younesi-Kordkheili and Pizzi [36] studied the effects of Soda and Kraft lignin on the properties of LUF and LPF wood adhesive. Based on FTIR spectra and DSC analysis of lignin, Soda lignin exhibited higher phenolic hydroxyl groups and lower glass transition temperature (Tg) value compared to Kraft lignin. Gouveia et al. [12] revealed that enzymatic modification of Eucalyptus Kraft Lignin (KL) has resulted in an increase of the carboxyl groups and a disappearance of the phenolic hydroxyls, leaving the aliphatic hydroxyl groups remained almost unchanged, conferring dimensional stability, hydrophobicity, and mechanical resistance. In this case, the laccase-assisted polymerization has potential as a green substitute for synthetic adhesives.

Lignin, a renewable resource, has potential to be used as green binders in the making of WPC, and hopefully will substitute the current use of petroleum-derived chemicals in the wood-based industry. Embracing the calls for a safer environment, utilization of lignin biopolymer as binder and additives in the making of WPC is in line to achieving the specific SDG targets. Circular Economy dictates the sustainable, rational, and efficient management of wastes and by-products because their annual volume remarkably increases and negative affects the environment. In contrast, waste recovery can provide a huge secondary resource for producing new materials and reducing their costs. Great attention must be drawn to the rational and effective utilization of both wood and polymer wastes, in this case kraft or soda-lignin or other abundant source of lignin biopolymer.

4 Protein-Based Binders for Wood-Plastic Composites

Protein-based binders derived from animal, shell, and chitosan sources have emerged as a promising solution for enhancing the adhesion between wood fibers and the matrix, while also providing eco-friendly benefits [38]. In this subchapter, we will explore the morphology, and physical and mechanical properties of protein-based binders for WPCs, with a focus on their application and impact on WPC composite performance. The morphology of protein-based binders in wood-plastic composites can affect the properties of the composites. Protein-based binders can take on various forms—powders, liquids or emulsions—and their morphology is a critical factor in influencing adhesion and strength of WPCs [3].

Particle size, shape, and distribution of the binders all play a role in determining their impact on the composite properties. For example, smaller-sized protein particles are able to penetrate and reach deeper into the spaces between wood fibers, resulting in better adhesion between the protein and the wood. A uniform distribution of the binders in the matrix can lead to a more homogeneous composite structure with improved properties. The physical properties of composites with protein-based binders depend on the nature of the binder and the process used to form the composite. The density of the composite, water absorption, and swelling properties are some of the physical properties that can be influenced by the protein-based binder.

The addition of protein-based binders can improve the physical properties of WPCs significantly. One of the key benefits of using protein-based binders is their water-repellent properties. This is because proteins can form a hydrophilic layer on the surface of the wood fibers, providing a protective barrier against moisture absorption [17]. Additionally, protein-based binders can improve the tensile, bending, and impact strength of WPCs. Studies have shown that composites with shell waste protein-based binders exhibit good water resistance and low water absorption compared to composites with other binders.

The mechanical properties of protein-based wood-plastic composites depend on various factors such as the nature of the binder, wood flour content, and the process used to manufacture the composite. Shell waste protein-based binders have also been reported to improve the mechanical properties of wood-plastic composites. The mechanical properties of WPCs such as stiffness, strength, and toughness can be significantly influenced by the addition of protein-based binders. Animal-based binders such as collagen and gelatin can improve the stiffness and strength of WPCs. The use of chitosan-based binders can also increase the mechanical properties of WPCs [8]. The addition of protein-based binders can also improve the impact resistance and fatigue properties of the WPCs. Studies have shown that composites with casein-based binders exhibit improved mechanical properties such as the flexural strength, tensile strength, and modulus compared to composites with other binders.

4.1 Application of Protein-Based Binders in WPCs

The use of protein-based binders in WPCs can be accomplished in various ways, depending on the intended application and desired properties. For example, animal-based binders can be attached to the surface of either the wood fibers or the plastic matrix before being cured. Shell-based binders can be used in conjunction with other additives to create a new composite material that has improved physical properties. The incorporation of protein-based binders can also depend on the production process used, as different methods can have varying effects on morphology, adhesion, and properties of the composite material. Protein-based binders have been shown to improve the properties of wood-plastic composites, and their use has been seen to be effective in the development of eco-friendly composites. The composites with protein-based binders exhibited good mechanical properties such as high stiffness, tensile strength, and flexural strength, as well as good water resistance, and low water absorption composites [18]. These properties could make the composite an excellent alternative to traditional materials in the construction industry. Protein-based binders from animal, shell, and chitosan sources are promising materials for enhancing WPC composite performance. Their morphology, and physical and mechanical properties make them unique among other binders—especially in their water repellency and improved adhesion to wood fibers. The use of protein-based binders presents an eco-friendly and sustainable solution for developing bio-adhesives for WPCs.

5 Hybrid Binders

Thermosetting resins (epoxy, phenolic, and polyester resins), thermoplastic resins (polyethylene, polypropylene, and polystyrene), natural binders (lignin, protein, and starch), and inorganic binders (cement and lime) are a few of the most common binders for WPCs, as mentioned previously. The choice of binder will be influenced by factors including the desired properties of the WPC, the intended application, and cost considerations. Researchers used to compare a thermosetting resin with a thermoplastic resin to solve the problem of particleboards being toxic and partly the problem of disposal of polymer waste. For example, Ibragimov and [15] conducted a study in which they examined the physical and mechanical properties of wood polymer composite materials that were based on secondary thermoplastic polyolefins and standard thermosetting binders.

Yet, some researchers take it a step further by combining two or more types of binders to achieve specific performance properties; however, no article has yet been published in Scopus databases with the exception of phenolic-thermoplastic, soy protein-isocyanate [28], and epoxy-polypropylene combination [30]. One of the primary reasons why researchers attempt to hybridize various types of binders is to solve for the binder's limitations in terms of its capabilities. For examples, [13] performed surface treatments on the surface of WPCs in order to chemically and/or physically modify the surface of WPCs in order to improve the adhesive bonding between WPCs formulated with polypropylene and a commercial epoxy adhesive. This was done in their study to improve the adhesive bonding between WPCs formulated with polypropylene and a commercial epoxy adhesive. Second, due to their hydrophilic nature, bio-based binders (starch, protein, and cellulose) cannot create an adhesive or operate as a binder alone [19].

Shi and Walker [28] list urea-based adhesives (like urea formaldehyde resin), phenolic-based adhesives (like phenol resorcinol adhesives), isocyanate-based adhesives, and adhesives made from renewable resources (like soybean, lignin) as examples of commonly used adhesives. Combining the renewable and biodegradable qualities of lignin with the enhanced mechanical properties of a polyurethane resin could be very useful. The following is an example of one possible option for a combination of hybrid binders. First, there's epoxy-polycarbonate, a hybrid binder combining the toughness and impact resistance of polycarbonate resin and the strength of epoxy resin. The second type of binder is a phenolic-thermoplastic hybrid, which has the advantages of both thermoplastic and phenolic resins. Third, lignin-polyurethane is a hybrid binder with increased mechanical capabilities, thanks to the polyurethane resin it incorporates while yet maintaining the renewable and biodegradable qualities of lignin. Fourth, epoxy-polypropylene is a hybrid binder that incorporates the best features of both epoxy resin and polypropylene resin, providing exceptional strength, durability, and flexibility as well as resistance to impact. Hybrid binder number five, soy protein-isocyanate, is a hybrid material that brings together the renewable and biodegradable qualities of soy protein with the enhanced mechanical capabilities of an isocyanate resin. Sixth, a hybrid binder made of cellulose acetate butyrate (CAB) and epoxy features the hardness and weather resistance of CAB resin along with the strength and durability of epoxy resin. Seventh is polyurethane-acrylic, a hybrid binder that features the best of both polyurethane and acrylic resins in terms of flexibility, chemical resistance, and weatherability/UV resistance. Table 2 summarizes the studies conducted on hybrid binders, with the epoxy-polypropylene combination drawing the most interest.

6 Recent Development in Eco-Friendly Binders

Eco-friendly binders are gaining attention due to increasing concerns over the negative impact of traditional binders on the environment. Traditional binders derived from non-renewable sources often contain toxic chemicals that can be harmful to human health and the environment. Other than binders that have been comprehensively discussed in this chapter, here are some recent developments in eco-friendly binders that are gaining popularity:

1. Soy-based binders: Soy-based adhesives have been developed as an alternative to petroleum-based adhesives. Soy-based adhesives are made from soy protein and can be used in a variety of applications, including wood composites, paper products, and packaging materials. Soy-based adhesives are biodegradable and renewable, making them an eco-friendly alternative to traditional adhesives.

Table 2 Studies conducted on hybrid binders

Authors	Descriptions
Rejeb et al. [25]	The primary goal of this research was to determine if a chromic treatment would enhance the adhesive coating of polypropylene (PP) WPCs. Three kinds of pulp fibers (kraft, thermomechanical (TMP), and chemothermomechanical (CTMP)) were used to strengthen PP at three different fiber contents (50, 60, and 70% w/w). A chromic treatment activated the surface of the PP-based WPCs and changed their roughness characteristics, providing a higher interfacial zone that increased epoxy coating adhesion
Rejeb et al. [26]	The primary purpose of this investigation was to investigate the effects of coating on the dimensional stability of WPCs. Kraft, TMP, and CTMP pulp fibers reinforced polypropylene and polylactic acid at 50, 60, and 70% w/w. Two commercial coatings, epoxy and acrylic, were used to cover the produced WPCs. The epoxy coating was more successful than the acrylic coating at reducing water absorption and thickness swell for all formulations
Shulga et al. [30]	The JSC "Latvijas Finieris" fine birch wood sawdust from plywood mechanical processing was chemically functionalized to create a novel lignocellulosic filler with improved adherence to recycled polypropylene for WPC. Alkaline hydrolysis and epoxy amine amination were used to functionalize wood sawdust. The mechanical characteristics (tensile, bending) of composites filled with original sawdust and laminated hydrolyzed microparticles improved with nitrogen content and decreased water sorption and swelling
Dimitriou et al. [6]	WPC components are hydrophobic and have low surface energy, making adhesion difficult. Epoxy glue bonding with hydrogen peroxide surface activation pretreatments improves lap-joint shear. This study suggests mild (pH 7.5) alkaline therapy. Free radical-based reaction reactions predominate in the acidic media, possibly altering the WPC material's polypropylene matrix
Gramlich et al. [13]	Surface treatments were used to chemically and/or physically change WPC made with polypropylene and commercial epoxy adhesive to increase adhesive adhesion. Profilometry and scanning electron microscopy of the water-treated WPC surface showed that water absorption and wood swelling increased bond strength by increasing bonding surface area. Flame and water treatments together increased shear strength, suggesting they may work together to strengthen adhesive connections
Dimitriou et al. [5]	Low surface energy and hydrophobicity of polyolefins, the most common polymer matrix, make WPC component adhesion difficult. This article examined four surface activation pretreatment methods for epoxy adhesive Bonding: hydrogen peroxide solution, hot air, gas flame, and halogen heating lamps. Increased bond strength induced material failure rather than bond line failure
Gill et al. [11]	In this study, melt blending extrusion was used to make PP-g-MA compatible PP-wood flour composites. The limited mobility of polymer chains caused by wood flour particles increased WPC stiffness

2. Biodegradable binders: Biodegradable binders are made from materials that break down naturally in the environment. One example of a biodegradable binder is polylactic acid (PLA), which is made from renewable resources such as cornstarch or sugarcane. PLA has been used as a binder in a variety of applications, including in the production of 3D printing filaments and packaging materials.

3. Cellulose-based binders: Cellulose-based binders are made from cellulose, which is a natural polymer found in plant cell walls. Cellulose-based binders are renewable, biodegradable, and non-toxic, making them an eco-friendly alternative to traditional binders. One example of a cellulose-based binder is carboxymethyl cellulose (CMC), which has been used as a binder in the production of ceramics and paper products.

References

1. Adekunle, K.F., Kumar, S., Banerjee, A.: Natural adhesives for wood and wood-based composites: a review. J. Adhes. Sci. Technol.Adhes. Sci. Technol. **34**(13), 1449–1469 (2020)
2. Bartolome, M.J., Satya, S., Padhi, P., Fichtberger, O.G., Schwaiger, N., Seidl, B., Kozich, M., Nyanhongo, G.S., Guebitz, G.M.: Improving properties of starch-based adhesives with carboxylic acids and enzymatically polymerized lignosulfonates (2022)
3. Bismarck, A., Mishra, S., Lampke, T.: Plant fibers as reinforcement for green composites. In: Polymer Composites, vol. 1. Macro-and Microcomposites–Structure-Property Relationships, pp. 25–54. Mohammad Luqman (2005)
4. Cornejo-Ramírez, Y.I., Martínez-Cruz, O., Del Toro-Sánchez, C.L., Wong-Corral, F.J., Borboa-Flores, J., Cinco-Moroyoqui, F.J.: The structural characteristics of starches and their functional properties. CYTA – J. Food **16**(1), 1003–1017 (2018). https://doi.org/10.1080/19476337.2018.1518343
5. Dimitriou, A., Hale, M.D., Spear, M.J.: The effect of four methods of surface activation for improved adhesion of wood polymer composites (WPCs). Int. J. Adhes. Adhes.Adhes. Adhes. **68**, 188–194 (2016)
6. Dimitriou, A., Hale, M.D., Spear, M.J.: The effect of pH on surface activation of wood polymer composites (WPCs) with hydrogen peroxide for improved adhesion. Int. J. Adhes. Adhes.Adhes. Adhes. **85**, 44–57 (2018)
7. Diéye, Y., Gueye, P.M., Toure, P.M., Bodian, S., Sambou, V., Tigampo, S.: Comparison of two types of binders naturals on the mechanical and thermal properties of typha leaf powder panels. Sigma J. Eng. Nat. Sci. **38**(4), 2069–2081 (2020)
8. Dong, Y., Zhang, M., Rong, M.Z., Han, X.J.: Water-resistant wheat gluten-based adhesives. Ind. Crops Prod. **52**, 33–37 (2014)
9. Flory, A.R., Vicuna Requesens, D., Devaiah, S.P., Teoh, K.T., Mansfield, S.D., Hood, E.E.: Development of a green binder system for paper products. BMC Biotechnol. **13** (2013)
10. Gadhave, R.V, Mahanwar, P.A., Gadekar, P.T.: Starch-based adhesives for wood/wood composite bonding: review 19–32 (2017). https://doi.org/10.4236/ojpchem.2017.72002
11. Gill, Y.Q., Abid, U., Irfan, M.S., Saeed, F., Shakoor, A., Firdaus, A.: Fabrication, characterization, and machining of polypropylene/wood flour composites. Arab. J. Sci. Eng. **47**(5), 5973–5983 (2022)
12. Gouveia, S., Otero, L.A., Fernández-Costas, C., Filgueira, D., Sanromán, Á., Moldes, D.: Green binder based on enzymatically polymerized eucalypt kraft lignin for fiberboard manufacturing: a preliminary study. Polymers **10**(6) (2018)
13. Gramlich, W.M., Gardner, D.J., Neivandt, D.J.: Surface treatments of wood-plastic composites (WPCs) to improve adhesion. J. Adhes. Sci. Technol.Adhes. Sci. Technol. **20**(16), 1873–1887 (2006)
14. Hegde, S., Thomas, S., Jose, J.: Bio-based polymers as a potential alternative for petroleum-based polymers in wood plastic composites. J. Polym. Environ.Polym. Environ. **28**(10), 2725–2739 (2020)
15. Ibragimov, A.M., Fedotov, A.A.: Research physic-mechanical properties of composite materials on the base of crushed wood (2018)

16. Jimenez, M., Schwaiger, N., Flicker, R., Seidl, B., Kozich, M., Nyanhongo, G.S., Guebitz, G.M.: Enzymatic synthesis of wet-resistant lignosulfonate-starch adhesives. New Biotechnol.Biotechnol. **69**, 49–54 (2022). https://doi.org/10.1016/j.nbt.2022.03.003
17. Kim, C.-H., Lee, H.-T., Kim, M.-J., Lee, D., Kim, H.-J., Kim, K.-M.: A comparative study on the properties of wood–plastic composites and traditional materials. BioResources **11**(3), 6275–6282 (2016)
18. Kržan, A., Simič, R.: Flexural properties of flat-pressed WPC panels reinforced with hexagonal pattern-shaped honeycombs. Wood Res. **60**(5) (2015)
19. Liew, K., Tan, Y., Albert, C.M., Raman, V., Boyou, M.: Potential of using natural and synthetic binder in wood composites. Forests **13**(6) (2022)
20. Liu, L., Qian, M., Song, P., Huang, G., Yu, Y., Fu, S.: Fabrication of green lignin-based flame retardants for enhancing the thermal and fire retardancy properties of polypropylene/wood composites. ACS Sustain. Chem. Eng. **4**(4), 2422–2431 (2016)
21. Oduola, M.K.: Effect of starch on the mechanical and rheological properties of polypropylene. Am. J. Chem. Eng. **3**(2), 1 (2015). https://doi.org/10.11648/j.ajche.s.2015030201.11
22. Ojogbo, E., Ogunsona, E.O., Mekonnen, T.H.: Chemical and physical modifications of starch for renewable polymeric materials. Mater. Today Sustain. **7–8**, 100028 (2020). https://doi.org/10.1016/j.mtsust.2019.100028
23. Pizzi, A., Papadopoulos, A.N., Policardi, F.: A review: wood composites and their polymer binders. Polymers **2020**(12), 1115 (2020)
24. Ramesh, M., Rajeshkumar, L., Sasikala, G., Balaji, D., Saravanakumar, A., Bhuvaneswari, V., Bhoopathi, R. (2022). A critical review on wood-based polymer composites: processing, properties, and prospects. Polymers **14**(3)
25. Rejeb, M., Koubaa, A., Elleuch, F., Godard, F., Migneault, S., Bouslimi, B., Mrad, H.: Effects of chromic treatment on the surface properties of polypropylene (Pp) wood composites. Coatings **11**(7) (2021)
26. Rejeb, M., Koubaa, A., Elleuch, F., Godard, F., Migneault, S., Khlif, M., Mrad, H.: Effects of coating on the dimensional stability of wood-polymer composites. Coatings **11**(6) (2021)
27. Serra-Parareda, F., Delgado-Aguilar, M., Espinach, F. X., Mutjé, P., Boufi, S., Tarrés, Q.: Sustainable plastic composites by polylactic acid-starch blends and bleached kraft hardwood fibers. Compos. Part B: Eng. **238** (2022). https://doi.org/10.1016/j.compositesb.2022.109901
28. Shi, S., Walker, J.C.F.: Wood-based composites: plywood and veneer-based products. In: Primary Wood Processing: Principles and Practice, vol. 9781402043932, pp. 391–426. Springer Netherlands (2006)
29. Shulga, G., Neiberte, B., Jaunslavietis, J., Verovkins, A., Vitolina, S., Shakels, V., Livcha, S.: Lignin-containing adhesion enhancer for wood-plastic composites. BioResources **16**(2), 2804–2823 (2021)
30. Shulga, G., Neiberte, B., Matvejs, J., Uzulis, S., Verovkins, A., Vitolina, S., Jaunslavietis, J., Brovkina, J., Livcha, S.: Tailored functionalization of waste hardwood lignocellulose for making recycled polymer-based composites. In: Paper Presented at the European Biomass Conference and Exhibition Proceedings (2017)
31. Shulga, G., Neiberte, B., Verovkins, A., Jaunslavietis, J., Shakels, V., Vitolina, S., Sedliačik, J.: Eco-friendly constituents for making wood-polymer composites (2016)
32. Trex Company: Sustainability (2021). https://www.trex.com/sustainability/
33. Watkins, D., Nuruddin, M., Hosur, M., Tcherbi-Narteh, A., Jeelani, S.: Extraction and characterization of lignin from different biomass resources. J. Mater. Res. Technol. **4**, 26–32 (2015)
34. Wei, T., Yang, X., Huo, Y., Jia, X., Shan, W., Wang, W.: Lignin/ammonium polyphosphate-modified wood flour/high-density polyethylene composites. Mater. Chem. Phys. **290** (2022)
35. Younesi-Kordkheili, H., Pizzi, A.: A comparison among lignin modification methods on the properties of lignin–phenol–formaldehyde resin as wood adhesive. Polymers **13**, 3502 (2021)
36. Younesi-Kordkheili, H., Pizzi, A.: Lignin-based wood adhesives: a comparison between the influence of soda and Kraft lignin. Int. J. Adhes. Adhes.Adhes. Adhes. **121**, 103312 (2023)

37. Younesi-Kordkheili, H., Pizzi, A.: Nanolignin, a coupling bio-agent for wood-plastic composites. J. Renew. Mater. **11**(5), 2075–2083 (2023)
38. Zhang, Z., Ma, X., Fang, Z.: A review of the synthesis and application of chitosan-based hydrogels. Polym. Int.. Int. **65**(6), 684–697 (2016)

Potential of Green Binders as Matrices in Composite Materials

Fatin Nur Azmina Mohd Fauzi, Azra Umairah Anuar, Noor Najmi Bonnia, Nur Izzah Atirah Mat Hussin, and Aein Afina Redzuan

Abstract Creating a "green composite binder material from industrial waste stabilisation" offers a more environmentally friendly approach for the benefit of future generations. Unlike conventional binders, which often rely on non-renewable materials like petroleum-based plastics and have adverse environmental effects, green binders contribute to a greener environment. They also have the potential to stimulate market demand for sustainable and eco-friendly products, thereby driving innovation within the binder industry. Various green binders have been created, such as using green polyol-based polyurethane for grouting substances and integrating liquid natural rubber to augment the durability of thermoset polymers. Analysis has shown that green polyol-based polyurethane's mechanical, physical, thermal, and morphological characteristics make it appropriate for use as grouting materials. Using liquid natural rubber as a toughening agent, combined with the insertion of carbon black filler and kenaf fibre as reinforcements, has dramatically enhanced rubber-toughened polyester composites' mechanical, thermal, and electrical characteristics. The findings emphasise the potential of green binders as alternatives to petroleum-based binders.

Keywords Green binder · Liquid natural rubber · Green polyol polyurethane · Composite · Properties

1 Introduction

Green binders are composed of renewable and biodegradable materials, such as recycled paper, cardboard, and plant-based materials. Green binders, also known as eco-friendly binders, have the potential to contribute significantly to sustainability and environmental protection [1]. Conventional binders usually comprise non-renewable materials, such as petroleum-based plastics, which have negative environmental

---

F. N. A. Mohd Fauzi · A. U. Anuar · N. N. Bonnia (✉) · N. I. A. Mat Hussin · A. A. Redzuan
School of Physics and Materials Studies, Faculty of Applied Sciences, Universiti Teknologi MARA Shah Alam, Shah Alam, Malaysia
e-mail: noornajmi@uitm.edu.my

S. S. Osman Al Edrus et al. (eds.), *Green Lignocellulosic-Based Panels*, Engineering Materials, https://doi.org/10.1007/978-981-96-3404-0_7

impacts during production, use, and disposal. Green binders also have the potential to create a positive impact on the market by driving demand for more sustainable and environmentally friendly products. This demand could lead to innovation in the binder industry, resulting in even more sustainable and eco-friendly products. Many types of green binders are currently being developed. An example is liquid natural rubber and green polyol-based polyurethane [2]. Liquid rubbers had been favoured as a thermoset modifier. The toughening mechanism and eventually the fracture toughness depend on the morphology aspect between the modifiers and the thermoset. Over past decades, most previous and well-established studies of thermosets have covered the toughening process mechanism of the epoxy resins modified with liquid rubbers. The toughening mechanism includes the void growths caused by the rubber particle pullouts and shear yielding of the matrix. It involves the dispersion of rubber particles in spheres that creates morphology with the continuous phase of epoxy, which is created from the phase separation.

Polyurethane possesses numerous advantageous characteristics for grouting applications, including its lightweight nature, high permeability, exceptional tensile strength, low conductivity, and outstanding thermal stability compared to alternative materials [3]. Polyurethane can be categorised into two main types: those derived from petrochemicals and those sourced from natural origins. The composition of polyurethane is determined by the variations in polyol utilised. Currently, the production of polyurethane heavily relies on polyether polyol derived from petrochemical sources, as indicated by [4], who found that petrochemical-based polyol constituted approximately 90% of industrial polyurethane production. This preference is attributed to the higher availability of polyether polyol compared to other forms of polyols in the market [5, 6]. Petrochemical polyols are extracted from lignite and crude oil. However, the extraction and synthesis processes for petrochemical-based polyols are complex and face several challenges, including rising costs, rapid depletion of resources, and the requirement for advanced processing systems [7]. Additionally, the extraction process for petrochemical-based polyol poses environmental risks due to the emission of harmful byproducts.

Therefore, this study developed a natural-based polyurethane that functions as an environmentally friendly binder and can potentially mitigate environmental issues in the grouting industry [8]. This is due to the extraction method of natural polyols, which is straightforward and comparatively low risk. Extraction or synthesis of natural-based polyols is possible from natural resources, including vegetable and wood oil. Vegetable oils of numerous varieties, including castor, rapeseed, palm, and sunflower, can be utilised as polyol. Castor oil, which is non-volatile and non-edible, can be utilised to produce polyurethane.

Furthermore, it is acknowledged as an outstanding vegetable oil because it can serve as a standalone polyol for producing polyurethane. The capability of this substance to act as a polyol and engage in a reaction with isocyanate to form polyurethane is attributed to the presence of hydroxyl functional groups in its chemical structure [9]. Castor oil typically contains three hydroxyl functional groups. Despite its relatively low hydroxyl number, castor oil can yield rigid polyurethane foam that meets grouting material standards due to its exceptional mechanical properties. Superior properties pertain to the polyurethane foam's ability to achieve high crosslinking, which can be accomplished by adjusting the ratio of isocyanate to hydroxyl groups (NCO:OH) [10]. As noted by [11], the NCO:OH ratio represents the proportional relationship between materials containing NCO and OH functional groups.

Natural polyols can be reacted with isocyanate and other additives in varying NCO:OH ratios to generate grouting materials that are environmentally friendly. Castor oil is a vegetable oil derived from the castor plant, an easily accessible resource [12]. Additionally, castor oil is an exceptional vegetable oil due to the presence of a hydroxyl functional group in its chemical composition. This facilitates the direct use of castor oil in producing castor-based polyurethane grout without requiring any modification technique [13]. The remarkable modification process of castor oil can potentially reduce the manufacturing expenses of polyurethane grout derived from natural sources. An additional benefit of employing castor oil as the primary polyol in producing castor-based polyurethane grout is that, compared to other crops, the castor plant requires little water, fertiliser, pesticides, and maintenance. This reduces cultivation costs for castor plants, as they are more straightforward to cultivate and maintain [14].

In addition, the manufacturing process of castor oil is environmentally friendly since it brings about a 61% decrease in the consumption of non-renewable energy sources, such as burning fossil fuels, and a 36% reduction in global warming due to the decreased amount of carbon dioxide released into the atmosphere [15]. Continuous cultivation of castor plants can effectively decrease carbon dioxide emissions by absorbing carbon dioxide from the atmosphere during their growth, hence mitigating global warming [16].

2 Composite

A composite material refers to the combination of two or more elements to create a material that possesses enhanced qualities compared to the individual components employed individually. The discontinuous phase of reinforcement and the continuous phase of the matrix are the two essential constituents of a composite [17]. Contradicting the principle of metallic alloy, each of the materials in the composite will maintain their physical, mechanical, and chemical properties. Composites offer the advantages of lower density than bulk materials, high strength and stiffness, and lighter weight for the final parts. With the help of ongoing research and development, advanced composite industries are growing rapidly and are varied. The fundamental cause of this phenomenon can be attributed to the distinctive characteristics and advantages of these materials compared to homogeneous metallic alloy structural materials like titanium, aluminium, and steel. Composites possess the qualities of being lightweight, having excellent performance, and exhibiting corrosion resistance. They are utilised in several industries, including maritime, automotive, sporting goods, aerospace, and, most recently, infrastructure. The careful selection of materials, techniques, and product designs can significantly contribute to achieving a cost-effective composite end product. The main disadvantage lies in composites' relatively high production costs [18].

2.1 Matrix

The composite's continuous phase, referred to as the matrix, can consist of ceramic, metal, polymer, or carbon materials [19]. Its main roles include maintaining proper alignment and shielding components from environmental factors and abrasion. Each matrix material exhibits unique characteristics, for example, polymers typically have lower stiffness and strength compared to metals and ceramics. Metals offer adequate strength and stiffness but are highly ductile, whereas ceramics are brittle despite their high strength and stiffness. A strong bond between the matrix and reinforcement is essential for efficient load transmission through shear loading at the interface. While polymer and metal matrix composites benefit from a solid interfacial bond, ceramic matrix composites often prioritise enhancing toughness over altering strength and stiffness, necessitating a lower interfacial bond strength [18]. Additionally, carbon serves as a matrix in applications requiring extreme temperatures, such as rocket nozzles and carbon–carbon brakes (Composite World 2014).

2.2 Reinforcement

The reinforcing phase serves to enhance the strength and stiffness of the composite. Typically, the aim is to employ a reinforcing material that is more intricate, robust, and stiff compared to the matrix. The primary categories of reinforcement are fibres, particles, and structural elements. Particles may vary in shape, such as platelets or spheres, and possess relatively uniform dimensions in all directions. Particulate composites might exhibit lower strength and stiffness compared to those with continuous fibres. Due to processing challenges and brittleness, particulate reinforcement is typically limited to filling around 40–50% of the volume. In fibre reinforcement, the crucial aspect is the length-to-diameter (l/d) ratio, also referred to as the aspect ratio. Discontinuous fibres have short aspect ratios with random orientations, whereas continuous fibres have long aspect ratios with preferred orientations. Examples of continuous reinforcement include unidirectional, helical winding, and woven cloth, while chopped fibres and random mats exemplify discontinuous reinforcement. Optimal loading for these reinforcement types typically ranges up to 70% to achieve the desired strength and stiffness [18]. Structural composites can be formed by combining both homogeneous and composite materials. The geometric arrangement of various structural elements is pivotal in determining the composite's properties. Laminar composites and sandwich panels are two common types of structural composites [20].

In general, the strength of the fibre depends on the diameter. As the diameter decreases, the strength increases so that the strength of the composite is compared to the bulky material. Apart from that, small-diameter fibre offers better flexibility and susceptibility towards fabrication. However, the cost of processing will usually increase as the fibre diameter decreases. The most common fibres include continuous and discontinuous synthetic rayon, nylon, glass, carbon, and aramid [21]. Continuous fibre composites are preferred for higher strength and stiffness applications where higher cost is tolerable, and discontinuous fibre composites are applicable in cost-driven but tolerable strength and stiffness applications [18].

2.3 Polymer

Naturally occurring polymers extracted from animals and plants have been used for centuries. The natural polymers include rubber, wool and cotton, wood, silk, and leather. Proteins, enzymes, cellulose, and starches are other natural polymers essential in plants and animals' physiological and biological processes. At the end of World War II, the revolution of the materials industry was set by the emergence of synthetic polymers. Not only can synthetic polymers have relatively low production costs, but they also offer superior properties and can replace metal and wood parts in several applications.

Similar to metal and ceramics, the structural elements of the polymers are somehow convolutedly related to their properties. The molecules in polymers are often called macromolecules due to their vast size compared to the hydrocarbon molecules (composed of hydrogen and carbon). The atoms within each molecule are bound covalently together. For chain polymers, a string of carbon atoms is the backbone of each chain. Usually, each carbon atom individually bonds to two neighbouring carbon atoms on either side. Each remaining valence electron for every carbon atom tends to bond with atoms or radicals adjacent to the chain. Thus, the term repeat units describes these long structural molecules that are well repeated throughout the chain. On the other hand, the small molecule from which a polymer is synthesised is called a monomer. The terms monomer and repeat unit have different meanings, but the term monomer or monomer unit is occasionally used instead of the more appropriate term repeat unit. While a "mer" is introduced from the Greek word "meros", which means part, the term polymer thus gives the meaning of "many mers".

One of the ways to classify a polymer is according to its end-use. The polymer types include the widely used plastics and elastomers (rubbers), fibres, adhesives, films, foam, and coatings. The majority of polymeric materials can be classified as plastic. By definition, plastics are materials that depend on structural rigidity to sustain the load for many application purposes. Plastic materials are usually divided into thermoplastic or thermoset and will have any degree of crystallinity and all possible molecular structures and configurations (linear, isotactic, branched, and others). Elastomeric materials, on the other hand, can be recognised by their rubbery elastic properties. The crosslinks within the polymer will bind and restore the chains to their original conformation before applying load. Therefore, they can spring back to the initial form elastically after some large deformation [20].

3 Green Binder

The history of green binders dates back to the early twentieth century when natural polymers such as cellulose, starch, and proteins were used as binders in various applications. However, with the advent of synthetic polymers in the mid-twentieth century, the use of natural binders declined. Synthetic polymers, particularly those derived from petrochemicals, were cheaper, more durable, and easier to produce than natural binders. Recently, interest in the green binder has grown due to increasing environmental concerns and the need for more sustainable materials. The desire to lessen the environmental impact of binder production and disposal has driven the development of green binders [22].

Various green binders have been developed recently, including binders made from recycled materials, plant-based materials, and bio-based polymers. These binders have been used in various applications, including paper products, building materials, and composite materials. Today, green binders continue to be an area of active research and development as scientists and engineers work to find new and innovative ways to use renewable and sustainable materials as binders in various applications. Green binders are essential to creating more sustainable and environmentally friendly products [23].

Liquid natural rubber (LNR) and green polyol-based polyurethane are examples of green binders. LNR is derived from the latex of rubber trees. It can be a binder, including adhesives, coatings, and sealants. Green polyol-based polyurethane (PU) is another example of a green binder. Green polyols are made from natural sources such as vegetable oils, soybeans, and corn. Green polyol-based PU is used in automotive parts, furniture, and insulation [24]. Both offer low environmental impact and can be a sustainable alternative to petroleum-based binders.

4 Rubber-Toughened Thermoset

There are continuous efforts to overcome the natural brittleness of the thermoset resins. In most cases, they are normally blended or reacted with various kinds of additives and modifiers to improve their toughness and fracture resistance by forming a second dispersed phase. Liquid rubbers are favoured as a thermoset modifier. The toughening mechanism and eventually the fracture toughness are very dependent on the morphology aspect between the modifiers and the thermoset. Over the past decades, most of the previous and well-established studies of thermosets cover on the toughening process mechanism of the epoxy resins modified with liquid rubbers. The toughening mechanism includes the void growths that are caused by the rubber particle pullouts and shear yielding of the matrix. It basically involves the dispersion of rubber particles in spheres that creates morphology with the continuous phase of epoxy, which is created from the phase separation [25]. Besides modifying them with liquid rubbers, the efforts continue by using rubber particles, solid particles, tougher thermoplastics, and through formation by interpenetrating polymer network (IPN) [26].

According to [25], studies on the toughening of other thermosets such as polyester (UPE) and divinyl esters resins (DVER) are not as progressive as the studies on the epoxies. From the previous studies, it is found that immiscible rubbers are effective in toughening the UPE and DVER crosslinked with styrene (S) and there are also cases where the two phases formed at the initial stage of the reaction. This condition is similar to the particulate morphologies between the rubber/epoxy mentioned. In their study, Auad et al. found that carboxyl terminated poly(butadiene-co-acrylonitrile) (CTBN), which is a common toughening agent for epoxy resins, is an almost unreactive rubber with the DVER and S comonomers. Some toughening of the DVER sample was accomplished with 5% of CTBN as the Kic value of the fracture toughness is 2.5 times higher compared to the neat DVER.

Saleh et al. [27] include the carboxyl–, epoxy–, and amine-terminated butadiene — acrylonitrile copolymers (CTBN, ETBN, and ATBN) are sometimes used as toughening agents and initially show miscibility with epoxy. However, despite this miscibility, the rubber phase tends to separate during polymerisation due to minimal compatibility with the matrix. Consequently, the toughness of epoxies can be enhanced by incorporating an optimal concentration of rubber without significantly compromising other properties. Three primary mechanisms of toughening are identified, which may operate independently or in combination to enhance toughness: crazing, shear banding, and elastic deformation of rubber particles.

5 Natural Rubber

Natural rubber (NR) is considered as a polymeric substance that has a high molecular weight. NR is the most demanding mixture production in the recent rubber industry. NR is originally obtained from the milky-like fluid or so-called NR latex of a tropical tree, *Hevea brasiliensis* having viscoelastic properties [28]. The basic structure of NR is cis-1,4 polyisoprene, with isoperene being a diene and 1, 4 addition giving a double bond in each of the isoperene units in the polymer. Therefore, natural rubber shows similar reactions with the unsaturated polymer [29]. Initially, the rubber in latex starts as a colloid dispersion state and undergoes precipitation through the addition of acetic or formic acid, followed by processes of washing, pressing, drying, and smoking. Prior to commercial packaging, natural rubber must adhere to a set of quality standards outlined in the "green book", which was endorsed by the International Federation of Rubber Producing Nations in 1960. This publication serves as a standard reference for classifying rubbers used in the industry. As per the Standard Malaysian Rubber (SMR-system) of 1965, rubber bales are categorised based on two significant characteristics: the initial basic material (including latex basis, waste coagulum basis, and latex with waste coagulum basis) and viscosity (stabilised viscosity and non-stabilised viscosity).

NR is being manufactured into a variety of products due to its exclusive combination of properties. Sixty per cent of the application of the automobile tyre and tube productions alone is consuming NR, while more than 300 of automobile components are made up of rubber. In the electrical production field, NR is widely used as insulators, specifically as high-voltage insulating matting, rubber insulating gloves, line hoses, and blankets. In addition, engineering products for uses in soil stabilisation, in vibration absorption, and in road making are being extensively developed. Butadiene-acrylic-nitril rubber (NBR), ethylene propylene (EPDM), and chloroprene rubber (CR) are examples of synthetic rubbers [30]. Table 1 shows several properties of NR and synthetic rubbers.

In a study conducted by [29], efforts were made to achieve optimised molecular configurations, vibrational frequencies, natural bonding orbital (NBO) analysis, and thermodynamic parameters for cis-1,4-polyisoprene. This was accomplished through density functional CAM-B3LYP/6-311++ G(d, p) and Hartree–Fock HF/6 311++ G(d, p) calculations. In particular, molecular geometries are characterised by bond lengths, bond angles, and bond dihedral angles. The bond length represents the average distance between the centres of two bonded atoms, while the bond angle is the angle formed between three atoms across at least two bonds. The dihedral angle arises between two planes, each containing three atoms. Through the application of CAM and HF calculation methods, Charles and Gunasekaran observed that bond lengths and bond angles exhibit significant similarity. Shorter bond lengths correspond to stronger bonds between atoms. Increased bond strength leads to greater

Table 1 Properties of NR and synthetic rubbers [30]

Type of rubber	Properties
Natural Rubber (NR)	High static tensile strength ranging from 15 to 22 MPa, along with high elongation between 600 and 900%, coupled with exceptional elasticity even at low temperatures (with minimal change up to −10 °C) characterise this material. However, it exhibits poor ozone and degradation stability. Nevertheless, its excellent crude adhesion enables good confection ability
Butadiene-acrylic-nitril Rubber (NBR)	Good resistance to mineral oils and aliphatic solvents, and as acrylicnitril content increases, their heat resistance increases but their cold resistance and elasticity decrease
Ethylene-propylene (EPDM)	Very good heat resistance (up to 150 °C)
Chloroprene rubber (CR)	Weather and ozone proof, medium heat resistance (90–100 °C), self-extinguish effect, its oxygen index is 40–50, and resists mineral oils

overlap, resulting in reduced distances between nuclei and shorter bond lengths. Across isoprene, the bond between C2 and C5 is the longest (1.497 (CAM method) and 1.502 (HF method)), indicating its relative weakness. Conversely, bonds such as C1-H6, C1-H7, C3-H8, C4-H9, and C4-H10 exhibit comparatively shorter lengths, signifying greater strength.

By conducting a chemical analysis the NR latex structure, it is known that cis-1,4-polyisoprene with a high degree of a long chain with branches of non-hydrocarbon groups along it, such as the aldehyde and hydroxyl groups. Aldehyde group concentration is varied functioning as the crosslinker between the chains. There are also other compounds such as proteins, fatty acids, carbohydrates, phospholipids, and inorganic salts until 5%. There is 0.95% of protein that 27.2% of is will stays in the rubber percentage upon centrifugation. These compounds also known as the non-rubber components may act as compatibiliser agents with hydrophilic fillers at filler–rubber interface and stabilise the latex in the water medium.

The atomic charges are also an important aspect of the NR molecular structure. It is determined that there are electrons lost from the hydrogen atoms. The three hydrogen atoms (H_{11}, H_{12}, H_{13}) are connected to the carbon C_5 making it becoming more electronegative. C_3 is bonded with one hydrogen atom making it less electronegative than C_4 and C_1 as they are bonded to two hydrogen atoms. C_2 is bonded with C_1. Overall, C_3 and C_5 are the most positive carbon atoms. Generally, a material with a weak polar covalent bond acts as a good insulator. The larger the electronegativity differences between two atoms, the more polar the atom will be. In isoperene molecule, all the C–C and C–H bonds except for C_5 relatively have a weak polar covalent bond because of the minimal differences of the electronegativities, thus giving the NR insulating properties [29].

6 Liquid Natural Rubber

Liquid natural rubber (LNR) is a modified form of natural rubber (NR) commonly utilised as plasticisers, compatibilisers, adhesives, and coatings. Initially derived from NR latex, LNR undergoes a degradation process to shorten the NR chains. Previous research by [31] has revealed that LNR shares the same microstructure as NR but differs in having shorter polyisoprene chains. Additionally, the molecular weight (Mw) of LNR typically ranges around 5000, whereas for NR, it is approximately 900,000. Depending on its molecular weight, LNR can act as a crosslinking agent or plasticiser due to active terminals such as –OH, –OOH, and –C=O. These active terminals facilitate bonding between the NR matrix and plastic particles by reacting with them, making LNR a desirable toughening agent. LNR is often dispersed into polymer matrices like polyester to enhance the toughness of cured polyester [31, 32].

Photo and oxidative degradations are among the well-known degradation methods which involve the chain scissoring. Each method has its specific limitation of chain scission. Furthermore, the method and the reagent used could affect the type of functional groups that will usually be produced at the end chains of the NR. Many studies before had reported about the degradation of NR to telechelic LNR (TLNR) through a redox reaction, ozonolysis, usual oxidation, photochemical oxidation, and several others. Examples of degradation reagents that had been used to prepare LNR or epoxidised LNR are periodic acid, phenylhydrazine-ferrous chloride, propanal, potassium persulfate, and ozone.

The recent study involves the effect of pH conditions towards the presence of functional groups on the LNR chains. H_2O_2 and $NaNO_2$ were used as reagents to prepare the LNR through oxidative degradation process. It was found that in an acidic medium (H_2O_2), hydroxyl groups appeared at the end of the LNR chains while in an alkali medium ($NaNO_2$), carbonyl groups formed at the LNR chains. Hydroxyl end groups are the results from the degradation reaction as the decomposition of peroxynitrite acid which produced hydroxyl radicals. On the other hand, the carbonyl end groups were yielded from the redox reaction which occurs in the alkaline medium. It was also found that the acidic medium is a better degradation medium than alkaline as it effectively produced LNR with lower molecular weight (Mn).

6.1 LNR-Toughened Composite

Hisham et al. [33] had compared several mechanical properties between neat epoxy, LNR-toughened polyester (2.5, 5 and 7.5%), and glass fibre (untreated and treated) reinforced LNR-toughened polyester. They found that rubber-toughened polyester with glass fibre composite gave the best properties and the optimum percentage of LNR to be used was 2.5%. Bonnia et al. [31] showed evidence of excellent bonding between kenaf fibres and the polyester matrix for both polyester with LNR (3wt%) and polyester without LNR composites. For composites without LNR, gaps existed between fibres and resin, prior to curing process which involved shrinking of polyester chains. On the other hand, for composites with LNR, the kenaf reinforcement seems to be well bonded with the matrix phase.

According to [32], the addition of 3phr of LNR to the epoxy matrix gave the best and the optimum value of mechanical properties such as the tensile strength, fracture toughness, and impact strength. The main explanation would be due to the increasing of the rubber particles size as the LNR loading increased and the mechanical properties would be decreased as well. Later, Bonnia et al. [31] expand the study of rubber-toughened polyester by using nanoclay (closite 30B) as filler. By using 3% of LNR, they found that the ductility of the polyester increased and the brittleness had lessened. The impact strength of the clay-filled polyester with LNR is higher compared to clay-filled polyester without the addition of LNR, with an optimum clay percentage of 6%. Flexural strength for toughened polyester composites is higher than nano-toughened composite. Rubber-toughened polyester-filled 4% of nanoclay has the highest flexural strength and modulus which are 70 and 7076.7 MPa. The intercalation and exfoliation of clay in the polyester caused the number of crosslinks in the composite to decrease which affected the increasing and decreasing strength of the composite.

Saleh et al. [27] conducted a study on the synthesis and characterisation of liquid natural rubber as an impact modifier for epoxy (EP) resin. This study found hydroxyl, carbonyl, and epoxy functional groups on the LNR chain. There were slight reductions in T_g, tensile and flexural strength when LNR was incorporated into the epoxy. However, there was a significant increase in the fracture toughness, and it became more apparent at higher testing speeds. Through the SEM analysis, it can be seen that the dispersed rubber particles play a role as stress concentrators during the fracture. This helps to delocalise the matrix's plastic deformation by surrounding it with the particles. Furthermore, stress whitening zones created by the rubber particles emerged at the fracture surface of the modified EP minimising the effect of the plastic deformation at the tip of the crack.

7 Polyurethane

Polyurethane is a polymer made up of organic units linked together by urethane [34]. Depending on its application, it can be characterised as thermoset and thermoplastic. They are made by combining a polyol with a diisocyanate in the presence of catalysts and additives. Both polyol and diisocyanate are available in various variations and ranges, allowing for a broader range of applications for the materials produced. Polyurethane was first discovered in the 1930s and is currently utilised in many items, including coatings, plastic components, seat cushions, and more [35].

The phase segregation in polyurethane has hard and soft segments. Choosing the suitable polyurethane composition is critical because polyurethane can be used in various applications. A polyol provides elasticity in the supple areas of the material [36]. Chain extenders and isocyanate are used in the hard segment to give strength and rigidity at crosslinking points. Over the past few decades, polyurethane has undergone significant advancements, largely owing to its favourable characteristics and numerous benefits. These properties allow for swift production and adjustment during sample preparation to meet specific requirements. Polyurethane exhibits outstanding bio-stability, high abrasion resistance, customisable mechanical strength, excellent chemical resistance, and biodegradability [37].

7.1 Cellular Structure and Rigidity of Polyurethane Foam

Polymer foams originate from the expansion of polymers, typically comprising two main phases: a solid phase and a gaseous phase produced from a reaction with a blowing agent [38]. These foams can exhibit either an open-cell or closed-cell structure. Closed-cell foam is characterised by an intact cell membrane surrounding each bubble, whereas open-cell foam features a ruptured cell membrane [39]. Rigid polyurethane foams are formed by the chemical reaction between polyols and isocyanates, forming a cellular structure. Two methods are employed in creating the cellular structure of polyurethane: reactive and unreactive. The reactive technique involves a chemical reaction between isocyanate and water, facilitated by a chemical blowing agent. This reaction yields gaseous carbon dioxide and the urea group [40]. Conversely, the unreactive approach entails the interaction between isocyanate and a physical blowing agent with low boiling points, such as pentane and hydrofluorocarbon. This interaction leads to the evaporation of the blowing agents due to the heat generated from the interaction between isocyanate and polyol, resulting in the formation of a cellular structure in the polyurethane [40]. The gas can permeate into bubbles within a polyurethane matrix through either expansion or the merging of two or more bubbles. Upon completion of expansion, the polyurethane foam undergoes a curing process, leading to its hardening and curing [39].

The vast majority of polyurethane foam manufactured in the industry utilises physical blowing agents throughout production, offering more significant benefits than chemical blowing agents. Using a chemical blowing agent during the manufacturing process of polyurethane foam has the potential to generate undesired chemical reactions, resulting in alterations to the chemical composition of the polyurethane. Consequently, this can negatively impact the quality and performance characteristics of the generated polyurethane. A study by Modesti et al. [41] investigated the impact of chemical and physical blowing agents on the properties of structural polyurethane foams. The study revealed that using a physical blowing agent yields superior mechanical qualities in polyurethane foam when compared to chemically blown foams. Consequently, the utilisation of a physical blowing agent is widespread in the manufacturing of polyurethane foam.

The cellular structure of polyurethane can take the form of either open-cell or closed-cell arrangements, impacting its flexibility and stiffness. Polyurethane foam comes in three main types: flexible foam, semi-rigid foam, and rigid foam [38]. Various factors such as the glass transition, chemical composition, degree of crosslinking, and level of crystallinity of the polymer foam contribute to the foam's characteristics [38]. Open-cell polymer foam, typically flexible or semi-rigid, finds applications in automotive seating, bedding, furniture, and more [38]. Conversely, closed-cell polymer foam tends to be rigid and is preferred for load-bearing structures due to its ability to withstand high pressure.

Rigid polyurethane foam is a very flexible group of rigid thermoset polymer foam [42]. Rigid polyurethane foams were invented by Otto Bayer and his colleagues in the early 1940s [42]. The initial production of stiff polyurethane foams occurred as aircraft components utilised during World War II. The first production of the foam involved the utilisation of toluene di-isocyanate (TDI) and polyesters as the primary components, with water serving as the blowing agents to create the foam's cellular structure [42]. Nevertheless, the formulation and manufacturing methods of polyurethane have evolved.

Rigid polyurethane foam has been extensively applied in numerous sectors, including aviation and automobile components, household appliances, and construction [43]. The high concentration of closed-cell foams, comprising over 90% of the polyurethane matrix, is responsible for the stiffness of polyurethane [40]. An experiment on rigid polyurethane grout under compression was performed by [44] where the results obtained showed that the rigid grout can withstand high compression loading, which improves grouting technology for infrastructure maintenance. Rigid foam polyurethane frequently exhibits low densities and exceptional mechanical qualities. The low densities of polyurethane foam provide benefits in grouting by allowing for a reduction in the load applied to existing structures.

7.2 *Fundamental Chemistry of Polyurethane Foam*

Polyurethane is a densely connected structure that cannot be extruded, formed by combining polyol and isocyanate through a chemical reaction [8]. Isocyanate refers to compounds containing a reactive NCO group, while polyol refers to compounds with an OH group in their chemical makeup. The NCO group reacts with hydrogen atoms from the polyol (OH group) to form the urethane bond [45]. This ongoing reaction results in a high molecular weight for the polyurethane. If an excess of the isocyanate group is present, further reactions may occur, leading to the formation of secondary products such as isocyanate and allophanate crosslinking [45]. Therefore, it's crucial to use the appropriate formulation when manufacturing polyurethane. The composition can be adjusted based on the equivalent ratio of isocyanate to polyol (NCO: OH ratio).

Aside from the covalent bonds that maintain the structure of the polyurethane molecule, the hydrogen bonding between molecular sites also plays a significant role in determining the properties and performance of polyurethane foams [45]. Additionally, [46] found that varying the NCO:OH ratio can result in different mechanical characteristics of the generated polyurethane. The selection of polyol and isocyanate types is also a crucial component that affects the physical and mechanical characteristics of the resulting polyurethane foam. The polyurethane foam can exhibit either flexibility or rigidity and can contain varying degrees of urethane crosslinking. According to Yu et al., polymers that are rigid or tough have a significant degree of crosslinking in their structure. Thus, it may be inferred that a high degree of crosslinking in polyurethane results in exceptional mechanical strength.

7.3 *Type of Polyols*

The characteristics of the polyurethane produced are determined by the type of polyol used. The variation in OH value and equivalent weight among various types of polyol is responsible for this. As a result, the ultimate properties of the polyurethane foam produced will be impacted by the urethane crosslinking that occurs between the polyol and isocyanate. Flexible polyurethane foam results from the use of polyol with a high molecular weight, whereas rigid polyurethane foam would be produced with a low molecular weight [47]. Polyurethane production involves the use of either petrochemical-based polyol or natural-based polyol. Both vegetable oils and petrochemicals are highly competitive in terms of pricing. However, it is imperative to minimise the number of chemical processes involved in order to achieve competitive pricing for producing polyols [48]. Currently, the majority of polyols used in the production of rigid polyurethane foam are derived from petrochemical sources. However, the manufacturing process of polyether polyol involves a complex procedure that generates hazardous waste, raising environmental concerns [7].

Petrochemical-based polyol originates from a chemical reaction involving an organic oxide derived from petrochemicals and an initiator substance containing two or more active hydrogen atoms [40]. Synthetic rigid polyurethane foam is produced by blending a low molecular weight petrochemical-based polyol with isocyanate through a chemical reaction. Petrochemical-derived polyols are produced by catalysing the addition of propylene oxide or ethylene oxide to initiators containing hydroxyl or amine groups. They can also be synthesised by combining di-acids with glycols through polyesterification. Polyols extended with propylene oxide or ethylene oxide are referred to as polyether polyols, while those produced through esterification are termed polyester polyols [40]. Propylene oxide and ethylene oxide are sourced from crude oils and coals. Crude oil undergoes petroleum cracking to obtain propylene, which is then converted into propylene oxide. Incorporating the initial polyol into propylene oxide leads to the formation of polyether polyol [49]. Various initiators such as triethanolamine, trimethylolpropane, glycerol, sorbitol, pentaerythritol, and sucrose can serve as catalysts for polyether polyol production [40]. Table 2 outlines the properties of polyols used as initiators in the development of rigid polyurethane foams using petrochemical-derived polyols.

Table 2 Properties of polyols used as starters for the production of petrochemical-based polyols in rigid polyurethane foam manufacturing

Polyol as starter	Formula	MW	f	Hydroxyl number, mg KOH/g
Glycerol	$HOCH(CH_2OH)$	92.10	3	1827.3
Trimethylol-propane	$CH_3CH_2C(CH_2OH)_3$	132.0	3	1275.0
Triethanolamine	$N(CH_2CH_2OH)_3$	149.19	3	1128.0
Pentaerythritol	$C(CH_2OH)_4$	136.0	4	1650.0
Dipentaerythritol	$(HOCH_2)_3CCH_2OCH_2C(CH_2OH)_3$	254.0	6	1325.19
Xylitol	$HOCH_2(CHOH)_3CH_2OH$	152.0	5	1845.39
Sorbitol	$HOCH_2(CHOH)4CH_2OH$	182.17	6	1847.7
Sucrose	$C_{12}H_{22}O_{11}$ 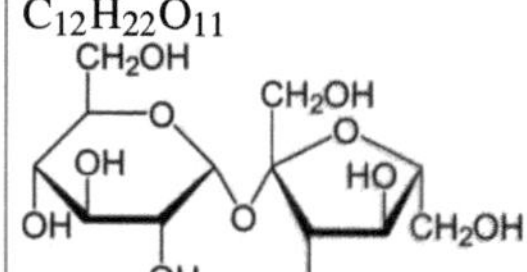	342.30	8	1311.1

A natural-based polyol is an alternative type of polyol that can be utilised in the manufacturing process of polyurethane. Natural-based polyol is derived from sustainable sources and subsequently synthesised into polyol [40]. Vegetable oils are natural-based polyols that can be utilised to produce polyurethane. Castor, palm, soybean, and linseed oil were extensively manufactured and utilised for many purposes [12, 50]. Vegetable oils were obtained from plants using many extraction methods. The majority of vegetable oils are comprised of unsaturated fats. Vegetable oils can undergo a chemical process where the double bond in the oil is changed into hydroxyl groups, creating polyols [40]. Several research have been conducted on developing rigid polyurethane foam using natural polyols such as soybean oil, palm oil, rapeseed oil and hydroxymethylated castor oil [51], 52. In previous research they developed HCFC and pentane-blown polyurethane rigid foams from modified soy-based polyol. However, the foam produced had inferior mechanical and thermal insulation properties. Several researchers also reported that most polyurethane foam produced from natural oil polyols was in a rigid or semi-rigid form where their rigidity was associated with the formation of crosslinked structures.

Castor oil is a single vegetable oil that can be utilised directly as a polyol without any modification. The hydroxyl functional group in its chemical structure allows it to react with isocyanate and generate urethane bonds. Castor oil has been significant in the polyurethane sector, even predating the availability of polyether polyol in the industry [40]. Castor oil production is a global industry, with an annual output of approximately 1,200,000–1,800,000 metric tonnes. Castor oil is derived from the seeds of the castor oil plant (*Ricinus communis* through a process called seed pressing. Castor oil is extensively utilised in several industries, particularly in polyurethane manufacture, due to the significant occurrence of a hydroxyl group known as ricinoleic acid.

Castor oil primarily contains 90% ricinoleic acid, 3% oleic acid, 1% stearic acid, 4% linoleic acid, and less than 1% linolenic acid. Its abundant ricinoleic acid content makes castor oil highly valuable and versatile in numerous sectors such as the chemical industry, medicine, cosmetics, paint, and various technological applications. With three hydroxyl groups per mole and a hydroxyl number of approximately 160 mg KOH/g [40], castor oil's hydroxyl functional group contributes to its resistance to oxidation. Typically, the ricinoleic fatty acid in castor oil consists of saturated or unsaturated chains with lengths of either 18 or 20 carbon atoms, and double bond unsaturation typically occurs at the 9th, 12th, and 15th carbon positions [53]. The extended carbon chain of castor oil enhances the thermal and hydrolytic stability of polyurethane synthesised with castor oil [54]. Being low in toxicity and readily available as a renewable agricultural resource, castor oil is highly suitable as a bio-renewable raw material for producing natural-based polyurethane.

7.4 *Isocyanates*

Isocyanate is an essential component in the production of polyurethane. The isocyanate supplies NCO groups to chemically react with OH groups from polyols and other components in the formulation [45]. Isocyanates can exist as either diol or polyfunctional compounds, with each molecule possessing two or more NCO functional groups. Approximately 61.3% of MDI isocyanates were utilised in the global market in 2000. TDI accounted for 34.1% of the total isocyanate use in the market, while the remaining 4.6% comprised various types of isocyanates. Additional isocyanates that can be found on the market include 1,6-hexane diisocyanate (HDI), isophorone diisocyanate (IPDI), 1,5-naphthalene diisocyanate (NDI), and 1,4-phenylene diisocyanate (PDI). These isocyanates are utilised in various applications. MDI is the most commonly utilised compound in manufacturing rigid polyurethane foam. The use of alternative isocyanates, such as TDI, in producing stiff polyurethane foam has several drawbacks, including excessive isocyanate content and high vapour pressure [42]. Thus, when it comes to producing stiff polyurethane foam, the MDI type of isocyanates is the most commonly employed [42].

There are three varieties of MDI accessible in the market: pure MDI, crude MDI, and polymeric MDI. Commercially available Pure MDI, also known as 4,4′ isomers, has two NCO groups per mole. The primary utilisation of pure MDI includes microcellular elastomer, polyurethane elastomer, and some flexible foams. Crude MDI refers to the 4,4′ MDI isomer, which has a high molecular weight and contains 2–3 NCO groups per mole. Its primary application lies in the manufacturing of extensively crosslinked polyurethanes, particularly rigid polyurethane foams. Polymeric MDI exhibits a high NCO functionality, with 3 NCO groups per mole. It is appropriate for use in the manufacturing of stiff polyurethane and polyisocyanurate [40]. Hence, for the creation of rigid polyurethane foam, which is a highly crosslinked polyurethane, the most appropriate types of MDI to utilise are either crude MDI or polymeric MDI.

MDI is synthesised through the catalysed reaction of phosgene with polyamines using aniline and formaldehyde as catalysts. Hydrogenating the respective nitro compounds results in diamino diphenylmethane or methylenedianiline (MDA), which is subsequently transformed into diphenylmethane diisocyanate (MDI). By regulating the condensation of aniline-formaldehyde, it is possible to modify the composition of isocyanates [42].

7.5 *Blowing Agents*

The establishment of the cellular, gas-filled structure of polyurethane occurs through a foaming reaction. Blowing agents are the substances employed in this process. There are two main types of blowing agents widely used: physical and chemical blowing agents. Chemical blowing agents are compounds that trigger gaseous transformations either within a system or through reactions with other components in the formulation [55]. One of the examples of chemical blowing agents is water. Research by Narine et al. [8] shows that the water-blown polyurethane rigid foams prepared from soy-based polyols had shallow closed-cell content. These resulted in inferior mechanical properties of rigid polyurethane foam produced. Therefore, alternative blowing agents, such as physical blowing agents, produce rigid polyurethane foam with excellent properties.

Physical blowing agents release gas through a physical process, such as evaporation at elevated temperatures, without undergoing chemical conversion [55]. Examples of physical blowing agents include chlorofluorocarbons and hydrochlorofluorocarbons. In the manufacturing of polyurethane foams, hydrochlorofluorocarbons, urethane-grade methylene chloride, and trichloroethane have been commonly used as blowing agents [45]. Hydrochlorofluorocarbons are particularly advantageous in polyurethane foam production due to their non-flammability, low toxicity, minimal diffusion rate, and lower thermal conductivity compared to air and other gases. Their low coefficient of heat conductivity enables precise control over the foaming process of polyurethane [45, 55].

7.6 Catalysts

The economic feasibility of polyurethanes depends on the effectiveness of a catalyst in regulating isocyanate reactions [42]. Two types of catalysts used in polyurethane manufacturing are amine compounds and organometallic compounds. Among these, tertiary amines like triethylenediamine (TEDA), dimethylcyclohexylamine (DMCHA), and dimethylethanolamine (DMEA) are widely employed in polyurethane synthesis [45]. They are chosen for their ability to catalyse urethane or gel formation, foam production, and isocyanate trimerisation. The free electron pair on the nitrogen atom of tertiary amines facilitates these catalytic reactions. The selection of an amine catalyst in polyurethane foam production can influence various aspects of the process, including cream time, rise time, and tack-free time [45].

7.7 Surfactant

The selection of surfactants during the production of rigid polyurethane foam plays a pivotal role in determining the foam's quality, including its fine, uniform, and closed-cell structures [42]. Surfactants, typically acting as stabilisers, are employed to modify the properties of polyurethane foams. These substances can take various forms such as polydimethylsiloxane–polyoxyethylene block copolymers, silicone oils, nonylphenol ethoxylates, and other chemical compounds [42]. Block copolymers are linked by either a hydrolytically stable Si–C bond or a less stable Si–C bond.

Surfactants in polyurethane foam emulsify the liquid components, regulate cell size, and stabilise the cell structure to prevent collapse and sub-surface cavities [42].

8 Properties of Rigid Polyurethane Foam

8.1 *Chemical Properties—Fourier Transform Infrared Spectroscopy*

Fourier transform infrared spectroscopy (FTIR) analysis relies on the interaction between infrared (IR) light and the organic resins present in a polyurethane sample. During FTIR analysis, IR radiation is directed towards a polyurethane sample. The resulting FTIR spectrum enables observation of the absorption or transmittance of IR light as it varies with wavelength. Each polyurethane sample exhibits a unique spectrum or pattern of absorption bands, determined by its specific chemical composition. The absorption of IR by the sample is influenced by various types of molecular motions, including vibrations, stretches, and bends, occurring within and between the chemical bonds of the coating molecules. According to [56], distinct absorption and transmittance patterns are exhibited by chemical groups such as aromatic, aliphatic, ester, amines, and isocyanate.

In their research, [57] produced a flexible and stretchable polyurethane grouting material. The FTIR spectra of the polyurethane grouting materials show a wide band stretching within the range of 3100–3600 cm^{-1}, suggesting an abundance of OH groups derived from the polyol used. The observed N–H stretching at 3378 cm^{-1} indicates the existence of the urethane linkage, thereby confirming the formation of the polyurethane grouting material. Carboxylic (C=O) stretching was observed at 1717 cm^{-1} which represents the presence of carboxylic in urethane linkage, while the band at 1091 cm^{-1} corresponds to the stretching of C–O of urethane linkage [57]. These confirm the urethane formation of the mixture.

8.2 *Thermal Properties—Thermogravimetric Analysis*

Due to the growing use of polyurethane foam as a thermal insulator and increasing in building fire break cases, these make thermal properties are one of the most trivial properties of polyurethane foam that need to be studied and taken into consideration depending on its final application [58]. Thermogravimetric analysis (TGA) is a method utilised for examining the thermal characteristics of a sample by monitoring its mass alterations during temperature variations. This analysis provides insights into various physical phenomena, encompassing phase transitions, absorption, and desorption, as well as chemical phenomena like chemisorption, thermal decomposition, and solid–gas reactions such as oxidation. TGA serves as a tool to evaluate the thermal endurance of a substance.

Jiao et al. [58] studied the thermal stability of rigid polyurethane foam produced from polyether polyol. Three stages of thermal degradation of rigid polyurethane foam were produced in their study. During the initial phase, there was a minor reduction in the mass of rigid polyurethane foam between temperatures of 110 and 190 °C, which was attributed to the breakdown of the urethane bond. The overall weight reduction equalled 2.7%, with the highest rate of weight loss occurring at a temperature of 150 °C. During the second stage, there was a significant drop in the mass of the grouting material between temperatures 210 and 397 °C. The overall weight reduction is equal to 46%, with the highest rate of weight loss occurring at a temperature of 314 °C. The reduction in mass was caused by the breakdown of isocyanate in the stiff polyurethane foam. In the final phase, there was a gradual decrease in the mass of the polyurethane grout observed between temperatures ranging from 397 to 670 °C. The most significant weight loss was recorded at 530 °C, resulting in a total weight reduction of 48.3%. This decline in mass during this phase was attributed to the decomposition of the polyol component [58].

8.3 Mechanical Properties—Compressive Strength

The compression test evaluates the ability of polyurethane grouts to withstand prolonged periods of compressive stress under a specified deflection. Sidek et al. [43] conducted a study investigating the influence of different ratios of polyol to isocyanate on the compression strength of polyurethane grout. The compressive strength of polyurethane was assessed at room temperature using a universal testing machine. The results showed the compressive strength graph of polyurethane grout with a steady increase trend associated with higher polyol-to-isocyanate ratios [43]. The polyol serves as an expansion agent, leading to an increase in the volume of polyurethane. Conversely, the inclusion of isocyanate in polyurethane production enhances the material's density and compressive strength, as isocyanate acts as a bonding agent in polyurethane synthesis [43]. Thus, their study demonstrated that an increase in the polyol–isocyanate mixing ratio resulted in improved mechanical properties of polyurethane.

8.4 Morphological Properties—Field Emission Scanning Electron Microscope

To assess the cell morphology of polyurethane foam, Field Emission Scanning Electron Microscope (FESEM) analysis was conducted. Rigid polyurethane foam is anticipated to exhibit a cellular structure primarily comprising closed-cell foams, typically spherical or polyhedral in shape. The dimensions, arrangement, and morphology of cellular foams within the polyurethane matrix have a substantial influence on the characteristics of the polyurethane material. The rigid polyurethane foams utilised in the study conducted by [59] are depicted in FESEM micrographs as having a polyhedral and spherical structure. An elevation in the isocyanate index led to a concurrent increase in the elasticity of the cell wall. Consequently, the expansion of cellular foams was impeded, leading to their reduction in size within the polyurethane matrix. As a result, the configuration of the cell foam became more compact, which enhanced the resulting polyurethane's compressive strength. These confirm that the mechanical properties achieved by polyurethane foam can be explained by observing their FESEM micrograph, which can show the cell morphology of polyurethane foam.

9 Conclusion

Green binders are becoming increasingly important in today's world due to the need for sustainable materials that reduce the environmental impact of human activities. Using green binders can help reduce the carbon footprint and improve the sustainability of various products while promoting renewable resource use. Green binders can improve the performance of materials and products, such as increased strength and durability. As recommendations, continued research and development are needed to identify new sources of green binders and to improve the properties and performance of existing ones. The research will help expand the range of applications for green binders and promote their wider adoption. Also, collaboration and partnerships among academia, industry, and government are needed to promote the development and adoption of green binders. This will help accelerate innovation, reduce costs, and promote the widespread adoption of green binders. Improving the development, adoption, and promotion of green binders can promote sustainability and reduce the environmental impact of materials and products.

Acknowledgements The author would like to thank the main supervisor for their guidance and intellectual in this research project. Lastly, thanks to the Faculty of Applied Science (FSG) for contributing to their laboratory and apparatus use.

References

1. Ghobadi, S., Şanlı, L.I., Bakhtiari, R., Gürsel, S.A.: Green composite papers via use of natural binders and graphene for PEM fuel cell electrodes. ACS Sustain. Chem. & Eng. **5**(9), 8407–8415 (2017)
2. Dîrloman, F.M., Toader, G., Rotariu, T., Ţigănescu, T., Ginghina, R., Petre, R., Alexe, F., Ungureanu, M., Rusen, E., Diacon, A., Ghebaur, A., Duldner, M., Coman, A., Ţincu, R.C.: Novel polyurethanes based on recycled polyethylene terephthalate: synthesis, characterization, and formulation of binders for environmentally responsible rocket propellants. Polymers **13**(21), 3828 (2021)
3. Zhang, Q., Hu, X.M., Wu, M.Y., Zhao, Y.Y., Yu, C.: Effects of different catalysts on the structure and properties of polyurethane/water glass grouting materials. J. Appl. Polym. Sci. **135**(27), 1–11 (2018)
4. Badri, K.H.: Biobased polyurethane from palm kernel oil-based polyol. Polyurethane 447–470 (2012)
5. Ali, A., Zafar, H., Zia, M., Ul Haq, I., Phull, A.R., Ali, J.S., Hussain, A.: Synthesis, characterization, applications, and challenges of iron oxide nanoparticles. Nanotechnol. Sci. Appl. **9**, 49–67 (2016)
6. Petrović, Z.S., Cevallos, M.J., Javni, I., Schaefer, D.W., Justice, R.: Soy-oil-based segmented polyurethanes. J. Polym. Sci. Part B: Polym. Phys. **43**(22), 3178–3190 (2005)
7. Adekunle, K.F.: A review of vegetable oil-based polymers: synthesis and applications. Open J. Polym. Chem. **5**, 34–40 (2015)
8. Narine, S.S., Kong, X.: Vegetable oils in production of polymers and plastics. In: Bailey's Industrial Oil and Fat Products, pp. 279–306 (2005)
9. Ibrahim Marhoon, I., Rasheed, A.K.: Mechanical and physical properties of glass wool-rigid polyurethane foam composites. Coll. Eng. J. (NUCEJ) **18**(1), 41–49 (2015). https://nahje.com/index.php/main/article/view/155/109
10. Oprea, S.: Effect of composition and hard-segment content on thermo-mechanical properties of cross-linked polyurethane copolymers. High Perform. Polym. **21**(3), 353–370 (2009)
11. Gogoi, R., Niyogi, U.K., Alam, M.S., Mehra, D.S.: Study of effect of NCO/OH molar ratio and molecular weight of polyol on the physico-mechanical properties of polyurethane plaster cast. World Appl. Sci. J. **21**(2), 276–283 (2013)
12. Datta, J., Głowińska, E.: Effect of hydroxylated soybean oil and bio-based propanediol on the structure and thermal properties of synthesized bio-polyurethanes. Ind. Crops Prod. **61**, 84–91 (2014)
13. Maisonneuve, L., Chollet, G., Grau, E., Cramail, H.: Vegetable oils: a source of polyols for polyurethane materials. OCL - Oilseeds Fats Crops Lipids **23**(5) (2016)
14. Biermann, U., Bornscheuer, U., Meier, M.A.R., Metzger, J.O., Schäfer, H.J.: Oils and fats as renewable raw materials in chemistry. Angew. Chem. – Int. Ed. **50**(17), 3854–3871 (2011). https://doi.org/10.1002/anie.201002767
15. Banik, I., Sain, M.M.: Water blown soy polyol-based polyurethane foams of different rigidities. J. Reinf. Plast. Compos. **27**(4), 357–373 (2008)
16. Gu, R., Konar, S., Sain, M.: Preparation and characterization of sustainable polyurethane foams from soybean oils. JAOCS J. Am. Oil Chem.' Soc. **89**(11), 2103–2111 (2012)
17. Majeed, K., Jawaid, M., Hassan, A., Bakar, A.A., Khalil, H.P.S.A., Salema, A.A., et al.: Potential materials for food packaging from nanoclay/natural fibres filled hybrid composites. Mater. Des. **46**, 391–410 (2013)
18. Campbell, F.C.: Structural composite materials. ASM Int. (2010)
19. Peters, S.T.: Handbook of Composites, 2nd edn. International Thomson Publishing (1998)
20. Callister, W.D.: Materials Science and Engineering: An Introduction, 7th edn. John Wiley & Sons (Asia) Pte Ltd. (2007)
21. De Albuquerque, A.C., Joseph, K., Hecker, L., Carvalho, D., Roberto, J.: Effect of wettability and ageing conditions on the physical and mechanical properties of uniaxially oriented jute-roving-reinforced polyester composites. Compos. Sci. Technol. **60**, 833–844 (2000)

22. Platania, E., Streeton, N.L.W., Lluveras-Tenorio, A., Vila, A.P., Buti, D., Caruso, F., Kutzke, H., Karlsson, A., Colombini, M.P., Uggerud, E.: Identification of green pigments and binders in late medieval painted wings from Norwegian churches. Microchem. J. **156**, 104811 (2020)
23. Das, O., Babu, K., Shanmugam, V., Sykam, K., Tebyetekerwa, M., Khorasani, S.N., Försth, M., Sas, G., Gonzalez-Libreros, J., Capezza, A.J., Hedenqvist, M.S., Berto, F., Ramakrishna, S.: Natural and industrial wastes for sustainable and renewable polymer composites. Renew. Sustain. Energy Rev. **158**, 112054 (2022)
24. Laurichesse, S., Avérous, L.: Chemical modification of lignins: towards biobased polymers. Prog. Polym. Sci. **39**(7), 1266–1290 (2014)
25. Auad, M.L., Frontini, P.M., Borrajo, J., Aranguren, M.I.: Liquid rubber modified vinyl ester resins: fracture and mechanical behavior. Polymer **42**, 3723–3730 (2001)
26. Bakar, M., Jaider, F.D.: Effect of plasticizers content on the mechanical properties of unsaturated polyester resin. J. Thermoplast. Compos. Mater. **20**, 53–64 (2007)
27. Saleh, A.B.B., Ishak, Z.A.M., Hashim, A.S., Kamil, W.A., Ishiaku, U.S.: Synthesis and characterization of liquid natural rubber as impact modifier for epoxy resin. Phys. Procedia **55**, 129–137 (2014)
28. Herculano, R.D., Tzu, L.C., Silva, C.P., Brunello, C.A., Queiroz, A.A.A., Kinoshita, A., et al.: Nitric oxide release using natural rubber latex as matrix. Mater. Res. **14**(3), 355–359 (2011)
29. Charles, J., Gunasekaran, S.: Experimental and theoretical investigations of natural rubber (cis-1,4 polyisoprene) using Coloumb attenuating and Hartree-Fock theoretical methods. Optik **127**, 279–287 (2016)
30. Renner, T., Pék, L.: Comparing strength properties of natural and synthetic rubber mixtures. Sustain. Constr. Des. (2011)
31. Bonnia, N.N., Ahmad, S.H., Zainol, I., Mamun, A.A., Beg, M.D.H., Bledzki, A.K.: Mechanical properties and environmental stress cracking resistance of rubber toughened polyester/kenaf composite. eXPRESS Polym. Lett. **4**(2), 4–10 (2010)
32. Lee, Y.P., Ahmad, H.S., Rasid, R., Se, Y.E.N., Yew, H.K., Tarawneh, M.A.: Effects of Liquid Natural Rubber (LNR) on the mechanical properties of LNR toughened epoxy composite. Sains Malaysiana **40**, 679–683 (2011)
33. Hisham, S.F., Ahmad, I., Daik, R., Ramli, A.: Penyediaan dan pencirian komposit UPR/Gentian kaca terawat menggunakan resin poliester tak tepu daripada bahan buangan PET, Prosiding Seminar Kimia Bersama UKM-ITB VIII (2009)
34. AlMaadeed, M.A.A., Ponnamma, D., El-Samak, A.A.: Polymers to improve the world and lifestyle: physical, mechanical, and chemical needs. Polym. Sci. Innov. Appl. 1–19 (2020)
35. Furukawa, M.: Science and technology of polyurethanes was the polyurethane understood in how far? Nippon Gomu Kyokaishi **92**(1), 30–37 (2019)
36. Jiang, Y., Lawrence, M., Ansell, M.P., Hussain, A.: Cell wall microstructure, pore size distribution and absolute density of hemp shiv. R. Soc. Open Sci. **5**(4), 171945 (2018)
37. Zhu, M., Li, S., Sun, Q., Shi, B.: Enhanced mechanical property, chemical resistance and abrasion durability of waterborne polyurethane based coating by incorporating highly dispersed polyacrylic acid modified graphene oxide. Prog. Org. Coat. **170**, 106949 (2022)
38. Klempner, D., Frish, K.C.: Introduction. In: Klempner, D., Vahid, S. (eds.) Polymeric Foams an Foam Technology, 2nd edn. Hanser Publishers, Munich (1991)
39. Saudners, J.H., Klempner, D.: Fundamental of foam formation. In: Klempner, D., Vahid, S. (eds.) Polymeric Foams an Foam Technology, 2nd edn. Hanser Publishers, Munich (2004)
40. Ionescu, M.: Chemistry and Technology of Polyols for Polyurethane, vol. 2. Rapra Publishing (2016)
41. Modesti, M., Adriani, V., Simioni, F.: Chemical and physical blowing agents in structural polyurethane foams: simulation and characterization. Polym. Eng. Sci. **40**(9), 2046–2057 (2000)
42. Backus, J.K.: Rigi polyurethane foam. In: Klempner, D., Sendijarevic, V. (eds.) Polymeric foams an foam technology, 2nd edn., pp. 121–123. Hanser Publishers, Munich (2004)
43. Sidek, N., Bakar, I. A.A., Azman, A.A., Rahman, A.S.A., Austin, W.A.: Strength characteristic of polyurethane with variation of polyol to isocyanate mix ratio: a numerical analysis. In:

Proceedings - 2017 IEEE 2nd International Conference on Automatic Control and Intelligent Systems, I2CACIS 2017, pp. 31–34 (2017)
44. Wei, Y., Wang, F., Gao, X., Zhong, Y.: Microstructure and fatigue performance of polyurethane grout materials under compression. J. Mater. Civ. Eng. **29**(9), 1–8 (2017)
45. Herrington, R., Hock, K.: Flexible polyurethane foams. In Klempner, D., Sendijarevic, V. (eds.) Polymeric Foams an Foam Technology, 2nd edn. Hanser Publishers, Munich (2004)
46. Oppon, C., Hackney, P.M., Shyha, I., Birkett, M.: Effect of varying mixing ratios and pre-heat temperature on the mechanical properties of polyurethane (PU) foam. Procedia Eng. **132**, 701–708 (2015)
47. Shaalan, M., Saleh, M., El-Mahdy, M., El-Matbouli, M.: Recent progress in applications of nanoparticles in fish medicine: a review. Nanomed. Nanotechnol. Biol. Med. **12**, 701–710 (2016)
48. Efstathiou, K.: Synthesis and characterization of a Polyurethane Prepolymer for the development of a novel Acrylate-based polymer foam. Budapest University of Technology and Economics (BME), pp. 1–57 (2011)
49. Hocking, M.B.: Handbook of Chemical Technology and Pollution Control, 3rd edn. Elsevier Ltd. (2005)
50. Louwrier, A.: Industrial products-the return to carbohydrate-based industries. Biotechnol. Appl. Biochem. **27**(1), 1–8 (1998)
51. Chian, K.S., Gan, L.H.: Development of a rigid polyurethane foam from palm oil. J. Appl. Polym. Sci. **68**(3), 509–515 (1998)
52. Samat, N.A., Mohamed Jais, I.B., Lat, D.C., Baharom, B., Salleh, M.A.: Strength of palm kernel oil-based polyurethane foam/resin as alternative method for ground improvement. ARPN J. Eng. Appl. Sci. **12**(23), 6611–6615 (2017)
53. Yusuf, A.K., Mamza, P.A.P., Ahmed, A.S., Agunwa, U.: Physico-mechanical properties of rigid polyurethane foams synthesized from modified castor oil polyols. Int. J. Sci. Res. Publ. **6**(7), 548–556 (2016)
54. Javni, I., Petrović, Z.S., Guo, A., Fuller, R.: Thermal stability of polyurethanes based on vegetable oils. J. Appl. Polym. Sci. **77**(8), 1723–1734 (2000)
55. Shutov, F.A., Visco, D.: Blowing agents for polymeric foams. In: Klempner, D., Frish K.C. (eds.) Polymeric Foams an Foam Technology, 2nd edn.. Hanser Publishers, Munich (2004)
56. Anonymous: FTIR analysis (2011). www.raecoatingslab.ca/ftiranalysis-of-pretective-coatings. Accessed 10 Oct 2016
57. Yang, Z., Zhang, X., Liu, X., Guan, X., Zhang, C., Niu, Y.: Flexible and stretchable polyurethane/waterglass grouting material. Constr. Build. Mater. **138**, 240–246 (2017)
58. Jiao, L., Xiao, H., Wang, Q., Sun, J.: Thermal degradation characteristics of rigid polyurethane foam and the volatile products analysis with TG-FTIR-MS. Polym. Degrad. Stab. **98**(12), 2687–2696 (2013)
59. Lim, H., Kim, S.H., Kim, B.K.: Effects of silicon surfactant in rigid polyurethane foams. Express Polym Lett **2**(3), 194–200 (2008)

Development of Recycled Cotton with Corn Starch as Binder for Sound and Thermal Insulation Panel

Tasnim Zakepeli, Kamarul Aini Mohd Sari, and Fatimah Mohamed Yusop

Abstract Concern over the environment causes the construction industry to look forward to new alternative insulation material toward sustainable and green materials. Given that textiles are used as insulation material in the construction industry, the possibility of using recycled textiles is relevant. The objectives of this study are to investigate the sound absorption coefficient and thermal conductivity of the recycled cotton textile. Recycled cotton textiles were intermixed with corn starch as binder. Four samples are developed with various ratios of recycled textile and binder, and different cutting dimensions of recycled cotton are used. This study investigated the physical, sound, and thermal properties of the prepared sample. Water absorption and density test have been conducted to determine the physical properties, and the measurement of sound absorption coefficient (SAC) was done by using the Impedance Tube Method. Meanwhile, the thermal conductivity was measured using SOLTEQ® model HE110. Two samples with recycled textile and corn starch in a 1:1 ratio (Sample S4) and a 5:6 ratio (Sample S5) show the optimum result of the proportions. The densities for S4, S5, and gypsum are 555.77, 677.78, and 668 kg/m^3. The result shows that S5 is 1.5% denser than gypsum panels. Water absorption test Samples S4, S5, and gypsum percentages are 105.8, 79.6, and 10%, respectively. For SAC at 500 Hz, the values for S4, S5, and gypsum are 0.3230, 0.4786, and 0.0500, where Sample S5 outperformed gypsum by 90% in SAC at low frequencies. The highest values obtained of SAC at high frequency (1000–8000 Hz) for S4 and S5 are 0.2035 and 0.235. Lastly, the thermal conductivities for S4, S5, and gypsum are 0.083, 0.087, and 0.3, respectively. Compared with gypsum, S5 performs 72% better. This study shows that S5 is classified as absorbing and suitable for low frequency and obtains better results in thermal conductivity compared to gypsum. However, to improve the result in WA, suitable material (plastic) can be considered to mix with this mixture.

T. Zakepeli · K. A. M. Sari (✉) · F. M. Yusop
Department of Civil Engineering Technology, Faculty of Engineering Technology, Universiti Tun Hussein Onn Malaysia, Pagoh, Johor, Malaysia
e-mail: aini@uthm.edu.my

S. S. Osman Al Edrus et al. (eds.), *Green Lignocellulosic-Based Panels*, Engineering Materials, https://doi.org/10.1007/978-981-96-3404-0_8

1 Introduction

Climate change is one of the most significant challenges of our era. The impact is unprecedented in scale and constitutes global challenges. It inspires individuals to become aware and concerned about environmental issues and sustainability. Our lifestyle must be altered to safeguard our planet. This climatic change is also associated with the manner in which we manufacture or develop items. Drastic measures must be implemented to ensure the situation does not deteriorate further. This is the rationale for researchers seeking innovations in products, including the materials utilized for manufacturing insulating panels. Nevertheless, the commercial panel continues to utilize synthetic fibers due to their superior performance in the mid-high-frequency band. The synthetic fibers encompass glass wool, rock wool, slag wool, and asbestos [4].

Given that textiles are utilized as insulation materials in the construction industry, the potential for employing recovered textiles is pertinent. Consequently, there are researchers seeking to utilize textile waste as the primary material for panel production. Several researchers utilize cotton textiles, discarded polyester, knitted fabric, and wool fibers to make insulating panels. Additionally, there are many types of binders utilized in the production of insulation panels. It may include natural binders such as maize starch, tapioca starch, and protein binders. The synthetic binder comprises urea formaldehyde resin (UF), clear water-based adhesive, and polyurethane.

The parameters often utilized to quantify sound and temperature absorption are categorized into three types: physical properties, acoustic properties, and thermal properties. This thesis examines the possibilities of recycling textile waste to create value-added materials, specifically sound and thermal insulation panels. Consequently, the utilization of recycled textiles for building insulation offers numerous benefits. Utilizing recycled materials can diminish fiber production by extraction or agricultural methods, which significantly contribute to the carbon footprint in the garment sector, and can mitigate the environmental detrimental consequences associated with textile waste landfilling [3].

2 Material and Method

2.1 Starch Binder

Starch is a readily accessible and cost-effective biodegradable substance. Starch is a white, often tasteless and odorless powder, consisting of tiny crystals or lumps that vary accordingly on their plant origin. Available starches comprise maize starch, potato starch, sago, and tapioca. Starches are altered to enhance their stability under extreme physical circumstances, modify their texture, adjust gelatinization time, or tailor their properties for specific purposes [5].

2.2 Thermal Insulation Material

Non-woven fiber insulation materials have several microscopic voids with trapped air layers, effectively inhibiting convective heat transfer. The incorporation of air layers and increased thickness in multiple fiber webs, without a commensurate increase in weight, enhances the thermal insulation properties of these materials [3].

2.3 Sound Insulation Material

Porous textile fiber materials can absorb sound energy, making them effective sound-absorbing materials. The diameter, length, and uniformity of the fibers are significant criteria that can affect sound absorption. High-density textile fibers can enhance sound absorption in the mid- to high-frequency regions. Conversely, denser materials have less air permeability while demonstrating superior sound absorption capabilities. The non-woven structure provides effective sound absorption in the mid- and high-frequency ranges, but is less effective in the lower frequency range of 100–400 Hz [3].

2.4 Methodology

This project aims to generate many samples for evaluating physical tests, sound absorption coefficients, and thermal conductivity. Figure 1 illustrates the methodological framework for this study.

Six trial samples have been created and designated as S1, S2, S3, S4, S5, and S6. S1 and S2 are designed to evaluate the efficacy of maize starch as a binding agent in cotton textiles. Textiles have been hand-cut with scissors into various dimensions, as illustrated in Fig. 2 (for Sample S3) and Fig. 3 (for samples S4, S5, S6). After that the procedure for the generated sample is illustrated in Fig. 4, while the testing machines for SAC and thermal conductivity are depicted in Figs. 5 and 6.

Sample preparation was conducted at the Composite Engineering Technology Laboratory. Meanwhile, the samples were testing in Building Services Laboratory, located at UTHM Pagoh campus. Two samples (S4 and S5) measuring 30 cm × 30 cm × 0.5 cm are utilized for density and thermal conductivity testing. S4 and S5, each with a diameter of 2.8 cm and a thickness of 1 cm, are utilized for water absorption and SAC at elevated frequencies. Simultaneously, S3, S4, S5, and S6, possessing a diameter of 10 cm and a thickness of 1 cm, are employed for SAC at low frequency.

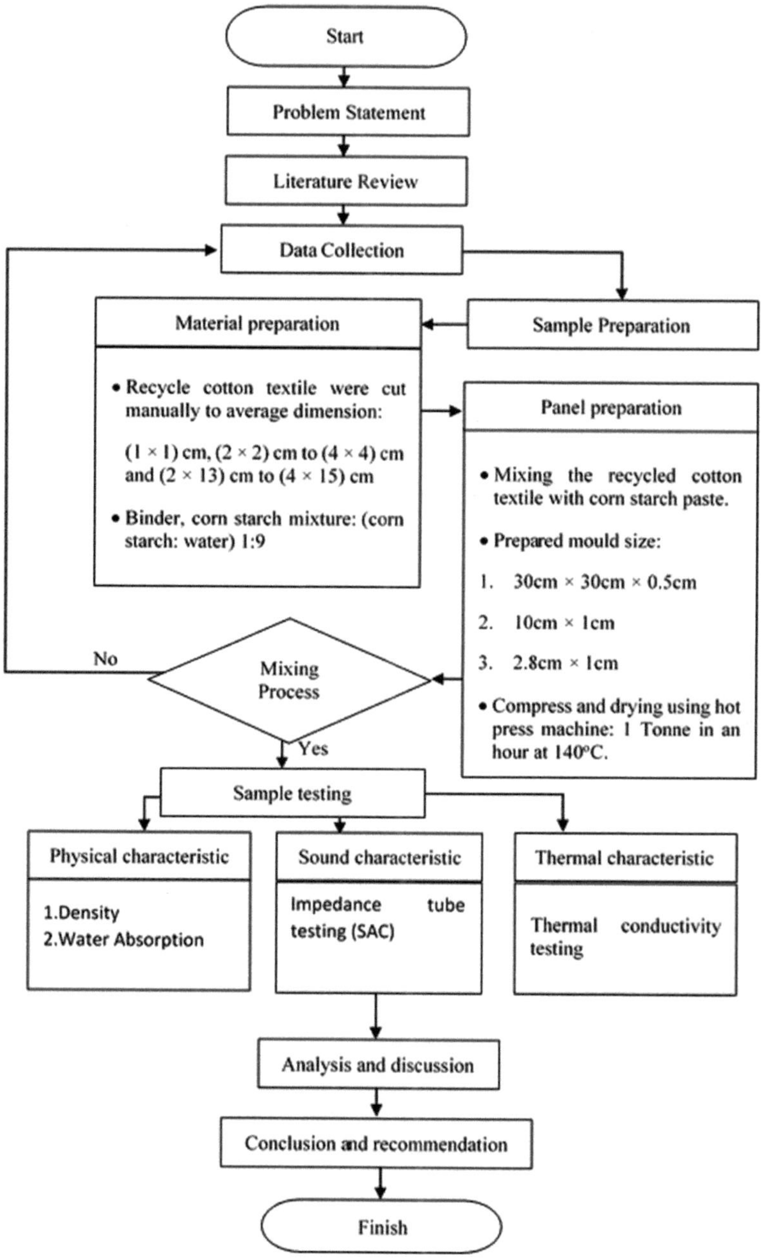

Fig. 1 Methodology flowchart

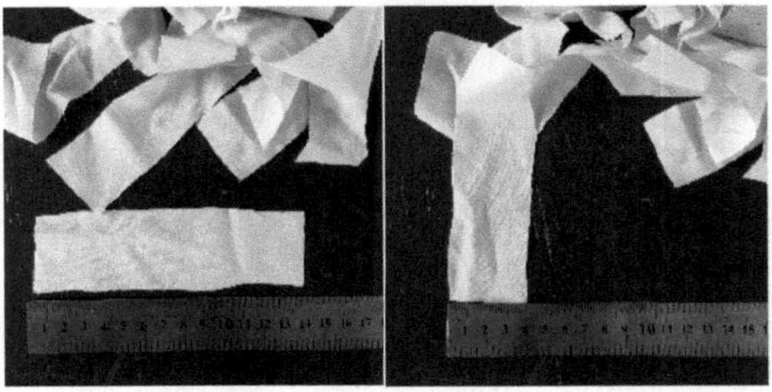

Fig. 2 Average dimension 2 cm × 2 cm to 4 cm × 4 cm

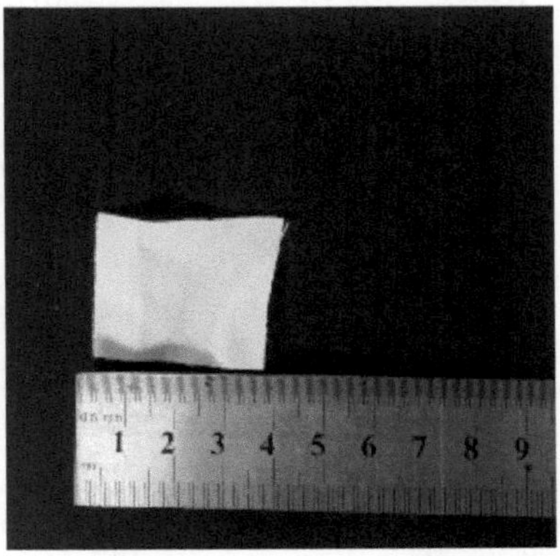

Fig. 3 Average dimension 2 cm × 13 cm to 4 cm × 15 cm

3 Result and Discussion

3.1 *Density, ρ*

The densities for samples S4 and S5, with dimensions 30 cm × 30 cm × 0.5 cm, have been determined. Sample S5 exhibits a greater density than Sample S4, as seen in Table 1. The density of Sample S5 is 122 kg/m^3 greater than that of S4 due to the extra 1 ratio of the corn starch in the sample.

3.2 *Water Absorption (WA)*

Samples S4 and S5, measuring 2.8 cm in diameter and 1 cm in thickness, are utilized to conduct a water absorption test. The initial masses for samples S4 and S5 are 5.2 and 5.4 g, respectively. Tests are administered for 2 h with a 10-min gap. Following the test, the percentage of water absorption for both samples is computed. Table 2 presents the water absorption % of the sample prior to and following a duration of 2 hours.

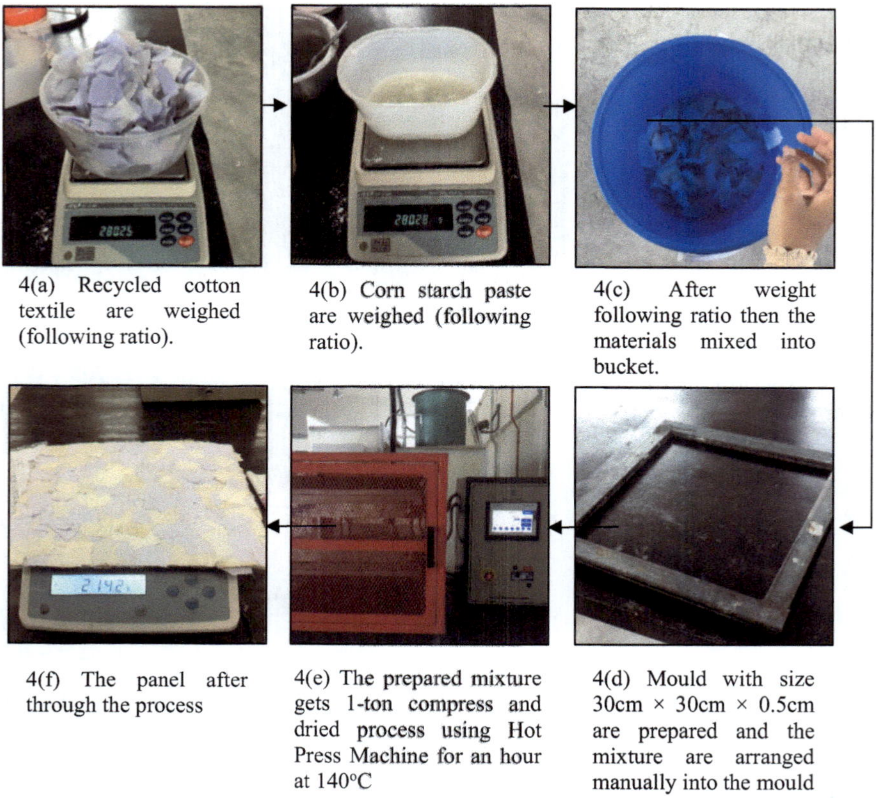

Fig. 4 Procedure for developing sample. **a** Recycled cotton textile are weighed (following ratio). **b** Corn starch paste are weighed (following ratio). **c** After weight following ratio then the materials mixed into bucket. **d** Mould with size 30 cm × 30 cm × 0.5 cm are prepared and the mixture are arranged manually into the mould. **e** The prepared mixture gets 1-ton compress and dried process using Hot Press Machine for an hour at 140 °C. **f** The panel after through the process

Fig. 5 Impedance tube for SAC test

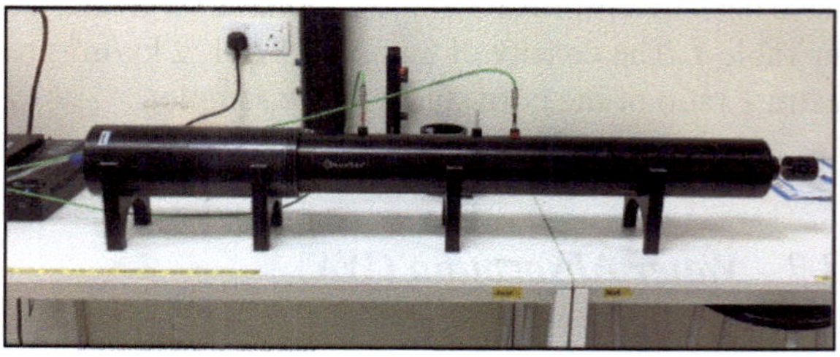

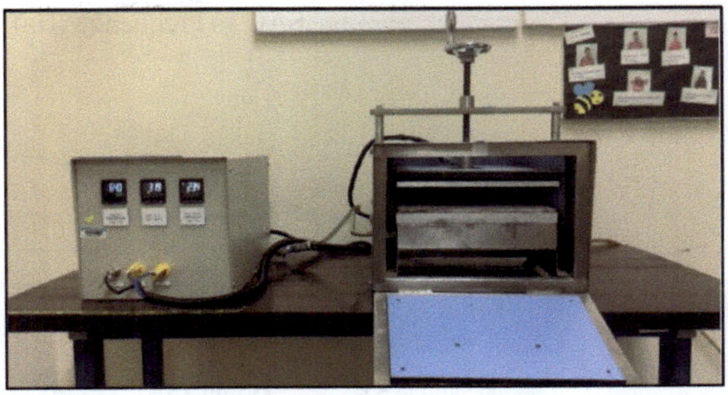

Fig. 6 Thermal conductivity machine SOLTEQ® model HE110

Table 1 Density of samples

Sample	Recycled cotton: Corn Starch	Mass of sample (kg)	Density, ρ (kg/m^3)
S4	1:1	0.250	555.77
S5	5:6	0.305	677.78

Table 2 Water absorption of samples

Sample	Initial mass (g)	After 2 h (g)	Water absorption (%)
S4	5.2	10.6	105.8
S5	5.4	9.7	79.6

According to Table 2, the water absorption percentages for S4 and S5 after 2 h are 105.8 and 79.6%, respectively. Sample S4 absorbed 26.2% more than Sample S5. Sample S5 has more density than Sample S4. Samples S4 and S5 have ceased to absorb water over the final 20 and 30 min of the test, respectively.

3.3 *Sound Absorption Coefficient (SAC)*

A sound absorption result was performed using an impedance tube to evaluate the sound absorption capacity of the generated samples. Four distinct samples (S3, S4, S5, and S6) are evaluated for low frequency, while two samples (S4 and S5) are assessed for high frequency. The samples are examined at low frequencies ranging from 125 to 1000 Hz. The samples are evaluated at high frequencies ranging from 1000 to 8000 Hz. Materials possess an exceptional capacity for sound absorption when the sound absorption coefficient approaches a value of 1 [1]. The sound absorption classification of composite samples was based on the standard ISO 11654. Figure 7 illustrates the sound absorption coefficient across all samples at low frequency in the 1/3 octave range. Figure 8 illustrates the sound absorption coefficient at the high-frequency 1/3 octave region for each sample, respectively.

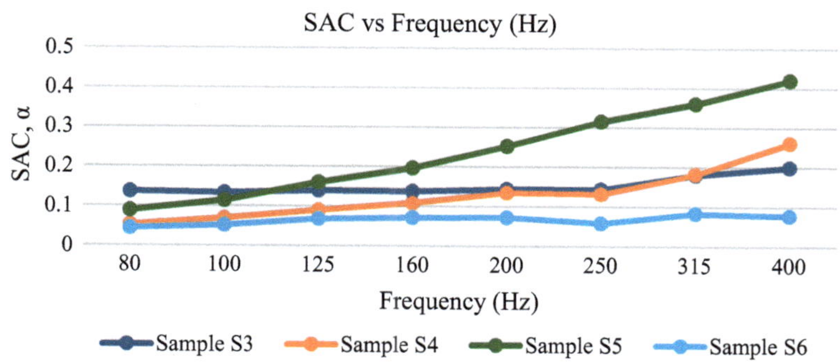

Fig. 7 Sound absorption coefficient at 1/3 octave band low frequency across all samples

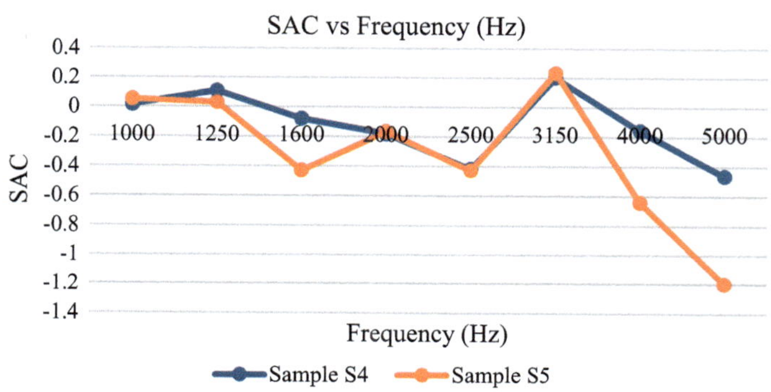

Fig. 8 Sound absorption coefficient at 1/3 octave band high frequency across all samples

Sample S5 (recycled cotton textile: corn starch, 1:1) has the highest sound absorption coefficient among the samples, with a value of 0.40 at a frequency of 400 Hz. Only Sample S5 may attain the absorption level. Sample S5 exhibits more density than S4 and S3; as the sample density increases, the sound absorption value rises in the mid- and upper frequency ranges [2].

The optimal SAC for Sample S4 in the 1/3 octave band at high frequency is 0.2038 at a frequency of 3150 Hz. The mean SAC for Sample S4 is −0.119. The optimal SAC for Sample S5 in the 1/3 octave band at high frequency is 0.2355 at a frequency of 3150 Hz. The mean SAC for Sample S4 is −0.3159. The optimal SAC indicates that the created sample is suitable as an insulating panel. Nevertheless, the graph depicted in Fig. 8 indicates that the data are unstable due to errors occurring during the insertion of the sample into the tube impedance sample holder.

Table 3 Thermal conductivity of the sample

Sample	Heat flow density q (W/m^2)	Hot plate temperature Th (^0C)	Cool plate temperature Tc (^0C)	Temperature difference ΔT (^0C)	Thermal conductivity k (W/mK)	Sample density, ρ (kg/m^3)
S5	31.0	49.8	30.0	19.8	0.087	677.78
S4	29.4	49.8	30.1	19.7	0.083	555.77

3.4 Thermal Conductivity

Table 3 shows the result of the samples for thermal conductivity. The thermal conductivity data indicate that Sample S5, composed of a recycled cotton to corn starch ratio of 5:6, had a thermal conductivity of 0.087 W/mK, surpassing Sample S4, which had a 1:1 ratio and a thermal conductivity of 0.083 W/mK. The study by [3] indicates that increased thermal conductivity signifies reduced resistance to heat transfer through the material. Sample S5 exhibits superior thermal conductivity relative to Sample S4 due to its greater density. This study demonstrates that an increase in density correlates with an increase in thermal conductivity.

4 Conclusion

This study has successfully achieved all objectives for an alternative of insulation panel made from recycled cotton textiles with maize starch as a binder. The experimental results indicate that Sample S5, with a ratio of 5:6, exhibits optimal density at 677.78 kg/m^3, superior water absorption, and a higher sound absorption coefficient at low frequencies compared to Sample S4. S5 attains a SAC of 0.4 at low frequency, outperforming gypsum by 90%. Comparing with Gypsum board, this sample with 70% has superior thermal conductivity compared to S5. Sample S5 exhibits superior performance in density and SAC at low density relative to gypsum. Gypsum exhibits superior performance in water absorption, specific acoustic conductance at high frequency, and thermal conductivity when compared to S5.

Acknowledgements The authors would also like to thank the Department of Civil Engineering Technology in Faculty of Engineering Technology, Universiti Tun Hussein Onn Malaysia, for its support and facility to conduct this study.

References

1. Borlea (Tiuc), A., Rusu, T., Vasile, O.: Investigation composite materials for its sound absorption properties. Romanian J. Acoust. Vib. **9**(2), 123–126 (2012)

2. Escobar, V.G., Gozalo, G.R., Pérez, C.J.: Variability and performance study of the sound absorption of used cigarette butts. Materials **12**(16) (2019). https://doi.org/10.3390/ma12162584
3. Hung Anh, L.D., Pásztory, Z.: An overview of factors influencing thermal conductivity of building insulation materials. J. Build. Eng. **44**, 102604 (2021). https://doi.org/10.1016/j.jobe.2021.102604
4. Islam, S., Bhat, G.: Environmentally-friendly thermal and acoustic insulation materials from recycled textiles. J. Environ. Manage. **251**, 109536 (2019). https://doi.org/10.1016/j.jenvman.2019.109536
5. Jhonson, A.C., Yunus, N.B.: particleboards from rice husk: a brief introduction to Renewable materials of Construction. Jurutera, 12–15 June 2009 (2009). http://ns41.webfirmdns.com/assets/download/Feature-Particleboard0609.pdf

Polyurethane from Liquefied Bamboo as Green Wood Coating Materials

Redzuan Mohammad Suffian James, Paridah Md Tahir, and Norwahyuni Mohd Yusof

Abstract Bamboo processing mills produce a substantial quantity of bamboo trash, which can reach up to 50% of the total bamboo utilized. Liquefaction presents an effective alternative for the utilization of bamboo wastes. Transforming this biomass into polyols by atmospheric liquefaction techniques, followed by its application in bio-polyurethane production, would significantly enhance the value of the bamboo business. Frequently employed liquefaction techniques may occur at elevated temperatures (150–250 °C) and atmospheric pressure, utilizing either acid or alkali catalysis, with the former being more prevalent. Numerous aspects must be considered in this process, with the most critical being the types of lignocelluloses, the ratios of the ingredients, and the temperature and duration of the reaction. The capacity of lignocellulosic biomass to diminish or substitute the extensive application of petroleum-derived liquefaction solvents in the market is anticipated to grow with improvements in liquefaction methods employing bio-based solvents. Although bio-polyurethane derived from biomass exhibits dark coloration, various ways have been documented to mitigate this problem. Contemplating the transformation of bamboo waste into light-hued polyols can then be employed in the manufacture of semi-transparent bio-polyurethane.

Keywords Bamboo · Polyol · Polyurethane · Coating

R. M. S. James (✉) · P. M. Tahir · N. M. Yusof
Institute of Tropical Forestry and Forest Products, Universiti Putra Malaysia (UPM), Serdang, Selangor, Malaysia
e-mail: redzuan.msj@gmail.com

P. M. Tahir
Faculty of Forestry and Environment, Universiti Putra Malaysia (UPM), Serdang, Selangor, Malaysia

S. S. Osman Al Edrus et al. (eds.), *Green Lignocellulosic-Based Panels*, Engineering Materials, https://doi.org/10.1007/978-981-96-3404-0_9

1 Introduction—Bamboo as Feedstock for Biochemicals

Predictions indicate that the rising demand for biomass in energy production will greatly diminishes wood resources. Exploring alternative materials is crucial. Kumar et al. [26] reported that global annual wood consumption is approximately 3.5 billion metric tonnes, showing a 65% rise since 1960. Cellulose, hemicellulose, and lignin are components of wood and other lignocellulosic materials. Cellulose is made up of insoluble ß-1,4-glucan fibres. Nanocellulose and other polysaccharides such as xylans, mannans, and glucans are examples of hemicellulose. Lignin, on the other hand, is a polyphenolic structure with a complicated structure. Wood from angiosperm trees normally comprises 42–50% cellulose, 25–30% hemicellulose, 20–25% lignin, and 5–8% extractives. This lignocellulosic pool is a key carbon sink in forest ecosystems, accounting for around 20% of terrestrial feedstock carbon storage and offering a massive, renewable source of feedstock for a variety of uses such as biofuels, adhesives, coatings, and so on. For example, only 10% of the oil palm is used as a final product, namely palm oil and kernel oil. Typically, 90% of the byproducts of palm oil production, such as empty fruit bunches, fibres, fronds, trunks, kernels, and palm oil mill effluent, are discarded as trash. Wheat straw, rice straw, palm, corncobs, corn stems, husk, and bamboo are only a few examples of lignocellulosic raw materials. The content and amount of cellulosic components vary in these materials. The yearly biomass output of cellulose is predicted to be 1.5 trillion tonnes by [26]. This substantial amount shows that cellulose may be regarded as a plentiful and sustainable raw material source for the creation of ecologically friendly and biocompatible goods. Improper waste disposal, such as open burning of crop leftovers, can result in the release of greenhouse gases such as carbon dioxide (CO_2), nitrous oxide (N_2O), and methane (CH_4). These emissions have the potential to endanger both human and natural habitats [23, 25].

Bamboo has become an important biomass resource worldwide in recent years [9]. The rapid economic growth in certain regions has contributed to making bamboo a feasible alternative resource. Researchers are interested in bamboo as a possible replacement for wood because wood resources are running low [36, 38]. Bamboo can be easily grown and used as a raw material for various beneficial purposes in the future. Wood biomass's efficiency for conversion is affected by specific gravity, chemical composition, fuel value, and size. Furthermore, the selected species for feedstock must have a high wood calorific value, the ability to produce wood with a high specific gravity, and the capability to grow rapidly in various environments. Sadiku et al. [36] state that the criteria mentioned have qualities that make bamboo a suitable option for substitution.

2 Deviation of Biochemicals from Biomass

Shen et al. [38] stated that plant biomass is a renewable resource that can be readily transformed into valuable compounds and fuels. Various studies have examined and listed the bio-based chemicals obtained from biomass, specifically bamboo, in Table 1.

Many scholars have extensively researched the liquefaction process of lignocellulosic biomass to produce liquefied products using polyhydric alcohol as a solvent under moderate conditions, as summarized in Table 2.

Various agricultural and forestry residue types are plentiful and renewable resources. This study specifically aims to investigate bamboo as a biomass resource. Sun et al. [42] proposed a new method for using bamboo in industry, which includes a liquefaction process using catalysts, solvents, and technology to create liquefied products. It is crucial to efficiently use leftover bamboo to address waste management and resource depletion issues. Lignocellulosic material, comprising cellulose, hemicellulose, and lignin, shows potential to replace petroleum-based polyols by converting it into reactive low-molecular compounds [28, 30, 46]. All products have been effectively transformed into polyols suitable for producing polyurethane materials with properties similar to traditional products.

At first, research on biomass liquefaction focused on liquefying altered feedstock materials. Due to the severe reaction conditions, intricate procedures, and high cost, researchers opted to directly convert biomass into liquid form and aimed to improve the process by changing the solvent and catalyst [52]. The liquefaction process consists of three main events: breakdown, esterification, and polycondensation, as indicated by previous studies [27, 52]. The reaction conditions consistently

Table 1 Methods used in the production of biochemicals from bamboo biomass

Biomass	Method	Product	References
Bamboo	Alkaline extraction and hydrolysis condition	Xylo-oligosaccharides (XOSx)	Song et al. [41]
Bamboo	Acidic steam explosion	Bioethanol	Wang et al. [47]
Hickory wood, bagasse, and bamboo	Pyrolysis temperature	Biochars and hydrochars	Sun et al. [43]
Bamboo	Liquefaction	Biopolyol and nanofibrils suspension	Xie et al. (2016)
Bamboo	Pyrolysis	Solid fuels (charcoals), liquid fuels, and gas (syngas)	Sharma et al. [37]
Bamboo	Enzymatic hydrolysis	Methane	Shen et al. [38]
Residual biomass, agricultural and municipal waste	Hydrothermal liquefaction	Biofuel	Dimitriadis and Bezergianni [11]

Table 2 Liquefied material and their target applications

Material	Application/Product	References
Algae	Biofuels	Chow et al. [8]
Bagasse	Biofuels	Zhang et al. [54]
White birch	Adhesive	Kobayashi et al. [22]
Bagasse and cotton stalk	Polyol	Hassan and Shukry [17]
Cornstalk	Polyol	Yan et al. [50]
Bamboo	Adhesive	Fu et al. [14], Sun et al. [42]
Cork powder	Polyol	Soares et al. [40]
Poplar wood	Polyurethane adhesive and surface finishing	Petrič and Sernek [35]
Black poplar	Polyurethane film	Budija et al. [3]
Empty Fruit Bunch (EFB)	Biofoam	Ali and Zubir [2]
Kenaf core	Polyurethane adhesive	Juhaida et al. [19]

influence the intensity of the reaction, the structure, and the performance of the products. Significant progress has been made by researchers in biomass liquefaction across different aspects such as raw materials, solvents, and catalysts.

Lignocellulosic materials are receiving considerable attention due to the perceived need to decrease dependence on fossil fuels, preserve the environment, and fulfil the requirements of different consumer products [3, 50]. These materials are widely regarded as the most plentiful renewable resources on a global scale. Lignocellulosic residue contains substantial quantities of cellulose, hemicellulose, and lignin, which are valuable for producing a variety of beneficial products [17].

Since 1925, when [13] first converted wood and plants into liquid, there have been many studies on the liquefaction of bioresources. In recent decades, many efforts have been made to improve the use of lignocellulosic biomass by applying chemical and biochemical techniques. The techniques mentioned are pyrolysis, modification, hydrolysis, and liquefaction [50]. Biomass can be converted into a liquid state by liquefying it in solvents at moderate temperatures and atmospheric pressure. This liquid substance has the potential to be used as biofuel, chemicals, and biomaterials, as emphasized by [54].

3 Factors Affecting the Efficiency of the Liquefaction of Lignocellulosic Biomass

Hu et al. [18] asserted that the efficacy of biomass liquefaction is influenced by the composition, structure, and morphology of lignocellulosic biomass materials. Recondensation reactions occur during biomass liquefaction, involving biomass derivatives and/or liquefaction solvents, competing with liquefaction reactions. Dominant

Table 3 Factors influencing the reaction efficiency during liquefaction of lignocelluloses [18]

Influencing factors	Function/Role
Type of lignocellulosic biomass	The composition, structures, and morphologies of lignocellulosic biomass materials impact the success of biomass liquefaction
Liquefaction solvent	Liquefaction solvents have critical roles in the liquefaction of biomass. To create polyol with appropriate properties for polyurethane (PU) applications, liquefaction solvents must not only permit effective and expeditious biomass liquefaction, but also display polyol attributes suitable for the desired PU applications
Catalyst	Lignocellulosic biomass liquefaction can be catalyzed by either an acid or a base; however acid-catalyzed liquefaction is more common. Concentrated (98%) sulphuric acid has the highest catalytic capacity in biomass liquefaction of all acids investigated to date, with typical loadings ranging from 1 to 3% (%wt. based on liquefaction solvent)
Liquefaction temperature and time	The liquefaction temperature varies greatly from 130 to 250 °C, according mostly to the type of catalysts used in the liquefaction process. Lignocellulosic biomass liquefaction often comprises a quick liquefaction stage within the first 15–30 min. Following this step, the liquefaction process proceeds at a much slower rate. The breakdown of more solvent-accessible biomass components such as lignin, hemicellulose, and amorphous cellulose dominated the initial rapid liquefaction stage, whereas the slow stage was dominated by the breakdown of densely packed and less solvent-accessible crystalline cellulose Acid-catalyzed liquefaction is normally carried out for 90 min at temperatures ranging 150–170 °C, ensuring high biomass liquefaction efficiency (>90% biomass conversion). Base-catalyzed liquefaction, on the other hand, necessitates higher temperatures of roughly 240–250 °C for 60 min in order to achieve comparable liquefaction efficiencies

recondensation reactions can decrease the efficiency of biomass liquefaction by increasing the amount of insoluble residue in polyols obtained from the process. This phenomenon has been well-documented in acid-catalyzed biomass liquefaction processes [29]. Research indicates that optimizing liquefaction parameters can effectively minimize or prevent the negative effects of recondensation reactions on biomass liquefaction. The parameters consist of liquefaction temperature and time, catalyst loading, and solvent-to-biomass weight ratio [12, 18, 20, 24, 27, 35], Li et al. (2017). Table 3 provides a detailed summary of the factors that impact the liquefaction process.

4 Chemical Composition of Lignocelluloses

Lignocellulosic biomass, consisting of cellulose, hemicellulose, and lignin, has gained significant attention as a feedstock for thermochemical conversion due to its high annual production and minimal emissions of harmful substances like NOx and

SOx [21]. The holocellulose component of biomass, consisting of cellulose and hemicellulose, typically makes up 43.7–77.6 wt.% of its chemical composition. Forestry and agricultural residues have lignin concentrations ranging from 12.0 to 53.5 wt.%. Softwood has a lignin concentration of 28.4–31.3 wt.%, hardwood ranges from 22.1 to 32.0 wt.%, and herbaceous crops have a lignin concentration of 13.3–24.8 wt.%. They constitute approximately 90% of the total mass of bamboo. Bamboo's chemical composition is similar to that of wood, but it contains additional minor components such as colours, tannins, protein, fat, pectin, and ash. The constituent is crucial for the physiological activity of bamboo and is located in cell cavities or specialized organelles.

5 Polyols

Polyols are chemicals characterized by many reactive hydroxyl groups in a single molecule. Polyols are classified into three categories based on their molecular weight: polyester polyols, polyether polyols, and polyols generated from polybutadiene. Polyether polyols differ in functionality, molecular weight, and hydrophobicity. This polyol is extensively employed in urethane adhesive applications owing to its combination of performance and cost-efficiency. Polyether polyols are the predominant category of polyols utilized in polyurethane synthesis, accounting for about 90% of the polyols accessible for acquisition. Polyethers often utilized possess molecular weights ranging from 500 to 2000 for diols and 250–3000 for triols. Polyether polyols comprise polyethylene glycol, polypropylene glycol, and diethylene glycol.

6 Polyurethanes

Since the groundbreaking research by Bayer and his associates in 1937, a variety of polyurethane compounds with distinct properties have been produced and extensively utilized as rigid and flexible materials. Polyurethane is, in numerous respects, distinct and more versatile than other contemporary polymers. Polyurethane has emerged as one of the most desirable polymers owing to its manufacturability and superior properties. It can be employed in rigid and flexible foams, elastomers, adhesives, coatings, resins, and numerous other applications. Rigid and flexible foams are prevalent and extensively utilized in the construction, transportation, bedding and furniture, and packaging industries. Polyurethane can be synthesized with or without isocyanates; however, the reaction involving isocyanates is predominantly utilized, especially in the commercial production of bio-based polyols and polyurethanes [29]. Polyurethane currently holds a significant market share in the plastics industry and in coatings, adhesives, sealants, and elastomers (CASE). Comprehensive research and ongoing advancements in formulations, additives, and processing techniques have led to a wide range of polyurethane applications [4]. The principal advantages of polyurethane are its durability, resilience, and outstanding chemical resistance.

Polyurethane distinguishes itself from most other polymers available in the market by enabling the processor or fabricator to modify the polymer according to specific requirements or to achieve particular final product characteristics [16]. The qualities of the final material are contingent upon the factors in polyurethane. Polyether polyols offer significantly enhanced flexibility in products, whilst polyester polyols deliver superior hardness and strength [1]. Isocyanates exhibit similar variations, as distinct types (aliphatic and aromatic) impart differing properties to the resultant polyurethane. The synthesis of polyurethane involves multiple components that influence its characteristics.

7 Polyurethanes from Liquefied Bamboo

Bamboo is employed for various applications, including flooring, panelling, furniture, fencing, engineered wood, and structural components. Presently, over 60% of bamboo is transformed into processing waste in modern bamboo farms and the associated enterprises. Bamboo waste has likely been discarded in landfills or incinerated, leading to resource depletion and environmental contamination. The transformation of bamboo into bio-based products can benefit agriculture, processing industries, and the environment. Table 4 summarizes the environmentally sustainable polyurethanes derived from bamboo-liquefied polyol compounds.

8 Polyurethane Coatings

Szycher [44] asserts that thin polyurethane surface films offer advantageous properties that exceed their minimal thickness. They can be easily applied using various techniques and can possess a broad spectrum of modulus as required. These films exhibit robust adhesion to numerous substrates. They are durable, impact-resistant, highly abrasion-resistant, and exhibit resistance to thermal ageing, aqueous solutions, fuels, oils, and various other chemicals. When aliphatic isocyanates are employed, they also demonstrate significant weathering resistance. Aliphatic-based polyurethane is increasingly utilized in corrosion-preventive maintenance coatings.

Polyurethane coatings are extensively utilized in diverse applications, including automotive, electronics, wood products, and machinery, due to their superior chemical, solvent, and abrasion resistance, alongside their toughness, favourable low-temperature performance, non-isocyanate reactivity, and one-component systems, with the first two properties being predominantly employed [44]. Two-component systems consist of a polyol component and an isocyanate component, which are blended before application and subsequently cured under ambient conditions. In two-component polyurethane coatings, oven-cured formulations can be achieved

Table 4 Some polyurethane products from liquefied bamboo

Species	Method	Product	References
Phyllostachys pubescens residues	Liquefied polyol mixed with polyaryl polymethylene isocyanate	Polyurethane foams	Gao et al. [15]
Bamboo leaves (the leaves and stems)	Liquefied polyol mixed with modified methylene diphenyl diisocyanate (MDI)	Polyurethane foams (acoustic material)	Chen and Jiang [5]
Bamboo powder	Liquefied polyol mixed with toluylene diisocyanate (TDI)	Semi-rigid polyurethane foam	Xu et al. [49]
Bamboo waste	Liquefied polyol mixed with methylene diphenyl diisocyanate (MDI)	Polyurethane foams	Zhang et al. [53]
Bamboo (epidermis and wax layers were removed)	Liquefied polyol (microwave assisted method) mixed with diphenyl diisocyanate (MDI)	Polyurethane foams	Xie et al. [48]

by combining a polyol with a blocked isocyanate prior to curing. Upon heating, the blocked isocyanate undergoes de-blocking, thereby exposing the isocyanate groups for reaction with the polyol. Non-isocyanate reactive solutions do not necessitate further reactions with isocyanates in applications like aqueous polyurethane dispersion. One-component systems comprise prepolymers with low isocyanate concentrations that undergo curing by a reaction with moisture. Consequently, they are frequently referred to as moisture-cured systems [44]. The morphological and thermal properties of polyurethane coatings are typically characterized by SEM, DMA, and/or DSC, whereas solvent resistance, viscosity, and tensile strength are generally assessed for coating applications.

The unique advantage of utilizing polyurethane is the extensive range of high-performance polymers that may be manufactured. Nevertheless, polyurethanes frequently contend with inexpensive polymers, as the overall expenses of the final product are influenced by factors beyond merely the pricing of raw materials. The considerations of equal significance encompass cycle time, tooling costs, finishing, rejection rates, and the potential for recycling.

At present, commercial polyol and polyurethane are predominantly synthesized from petrochemical derivatives [51]. Recent apprehensions around the exhaustion of fossil resources have generated significant interest in the development of bio-based polyol and polyurethane derived from renewable resources. Vegetable oils and lignocellulosic biomass are two primary categories of biomass that have been thoroughly investigated for their applications in the production of bio-based polyol and polyurethane.

The composition of soft and hard segments, the hydroxyl functionality of polyols, and the structure and concentration of hardeners are critical factors in polyurethane formulation. These elements significantly impact the physical, mechanical, thermal, and chemical properties of polyurethanes, consequently affecting their coating performance [32]. Physical, mechanical, thermal, and chemical properties of various potential applications of bio-based polyurethane coating have been summarized in Table 5.

9 Method to Produce Transparent/Translucent Colour Polyurethane

The colour of the liquefied biomass products is typically dark brown or nearly black. This can be a significant disadvantage when considering their use in industries that prioritize aesthetics, such as the coating industry [6]. It has recently been discovered that treatment with activated hydrogen peroxide can significantly lighten liquefied biomass without causing drastic degradation of the material. The only reports on light-coloured liquefied biomass-based finishes in the literature were reported by Cheumani-Yona in the years 2012 and 2015. The common liquefaction processes lead to dark-brown products, regardless of the raw materials used (such as sucrose, cellulose, wood). These products give the impression of degraded materials. The intense coloration is likely a result of the formation of carbonyl groups (C=O) and carbon–carbon double bonds (C=C) during the liquefaction process. The formation of molecules such as 5-hydrxymethylfurfural, 2-hydroxyethyl levulinate that undergo polymerization reaction to produce high molecular weight compounds have been, for example, reported for the liquefaction of cellulose, the major component of wood. The formation of char is also possible, particularly when the liquefaction is performed at higher temperatures (350–500 °C).

The method of bleaching polyol was introduced by Cheumani-Yona et al. [7] through the utilization of activated hydrogen peroxide. Some activators that have been introduced are Fenton reagent, Copper (II) chloride ($CuCl_2$) with pyridine, potassium carbonate, and Laccase/Mediator. Among these methods, the Cu(II)/pyridine/H_2O_2 and K_2CO_3/H_2O_2 systems exhibited higher efficiency compared to the traditional Fenton's reagent (Fe(II)/H_2O_2) although laccase appeared to be inefficient for the discoloration of liquefied wood. In general, it was shown in this work that the colour of liquefied wood can be changed by using activated hydrogen peroxide from dark brown/black to orange (mixture yellow and red) without drastic degradation of the material. According to the author, the described method could be generally applied also to other types of liquefied biomass and not only to liquefied wood. The same author, Cheumani-Yona et al. [6], carried out further analysis. They used potassium carbonate activated hydrogen peroxide to produce a yellowish polyol as a wood

Table 5 Various potential applications of bio-PU coatings

Composition	Finding	Potential applications	References
Vegetable oil-based Note Jatropha oil-based polyurethane acrylate (JPUA) Toluene diisocyanates (TDI) Isophorone diisocyanate (IPDI)	Viscosity: JPUA-TDI (10.820 Pas), JPUA-IPDI (0.096 Pas) GPC: JPUA-TDI (PDI 20.15), JPUA-IPDI (PDI 1.28) TGA: final residue JPUA-TDI 2.8 wt.%, JPUA-IPDI wt.1.2% Pendulum hardness: JPUA-TDI (118 s), JPUA-IPDI (90 s) Adhesion: brittle surface Contact angle: 75–89° Transmittance: transparent material (85%) Haze test: JPUA-TDI (22.3%), JPUA-IPDI (3%)	Clear coat coating such as for windows and for the automotive and construction industry	Mudri et al. [31]
Vegetable oil-based Note Hyperbranched polyurethane (HBPU) Linear polyurethane (LPU)	Solubility: HBPU are soluble in solvent 1,4-dioxane, acetone, ethyl acetate, whereas the LPU is not soluble. However, both types of polymers are soluble in xylene, DMF DMSO, and THF GPC: HBPU (PDI 2.3–2.8), LPU (PDI 4.8) Hardness: HBPU adequate flexibility (70–77 shore A), LPU (68 shore A) Impact: both HBPU and LPU have good impact (30–44 cm) Adhesion: both HBPU and LPU have good adhesive strength Tensile: HBPU (8.6 MPa and 721% elongation), LPU (5.31 MPa and 779% elongation) TGA: final residue HBPU 2–5 wt.%, LPU wt.2% Chemical resistance: Both HBPU and LPU are stable under chemical media such as aqueous 10% NaCl, 20% EtOH, 5% HCl, except for 3% NaOH	High-performance coating materials	Deka and Karak [10]

(continued)

Table 5 (continued)

Composition	Finding	Potential applications	References
Vegetable oil-based Note Polyurethane coating castor oil (PU-COs) Polyurethane coating hyperbranched polyol (PU-COHBO) Methylene diphenyl diisocyanate (MDI) Toluene diisocyanate (TDI) Isophorone diisocyanate (IPDI)	Gloss at 60°: PU-CO-MDI (102.7), PU-CO-TDI (110.5), PU-CO-IPDI (125.3), PU-COHBP-MDI (110.6), PU-COHBP-TDI (110.1), PU-COHBP-IPDI (140.0) Flexibility: all PU-Cos and PU-COHBPs have pass Pencil hardness: PU-CO-MDI (2H), PU-CO-TDI (2H), PU-CO-IPDI (3H), PU-COHBP-MDI (7H), PU-COHBP-TDI (7H), PU-COHBP-IPDI (6H) Cross-cut adhesion: PU-CO-MDI (2B), PU-CO-TDI (21), PU-CO-IPDI (2B), PU-COHBP-MDI (5B), PU-COHBP-TDI (5B), PU-COHBP-IPDI (5B) Impact resistance (lb. inch): PU-CO-MDI (74.25), PU-CO-TDI (67.50), PU-CO-IPDI (94.5), PU-COHBP-MDI (135), PU-COHBP-TDI (135), PU-COHBP-IPDI (135) Mar resistance (g): PU-CO-MDI (200), PU-CO-TDI (100), PU-CO-IPDI (300), PU-COHBP-MDI (425), PU-COHBP-TDI (550), PU-COHBP-IPDI (500) Chemical resistance Gel content (%): PU-CO-MDI (77.3), PU-CO-TDI (63.8), PU-CO-IPDI (70.2). PU-COHBP-MDI (92.8), PU-COHBP-TDI (80.3), PU-COHBP-IPDI (90.4) Electrochemical: Corrosion rate (blank mild steel panel > PU-COs > PU-COHBPs) TGA: final residue PU-COs and PU-COHBPs range between 1–5 wt.%	Industrial coating	Patil et al. [33]

(continued)

Table 5 (continued)

Composition	Finding	Potential applications	References
Biomass-derived alkylphenol polyol Note Bio-PU 2EH pinewood (BPUI) Bio-PU 2EH Stipa (BPUII) Bio-PU PG pinewood (BPUIII) 2-ethyl hexanol (2EH) Propylene glycol (PG)	TGA: final residue BPUI (23 wt.%), BPUII (19 wt.%), BPUIII (28 wt.%) Optical contact angle: BPUI (79), BPUII (86), and BPUIII (68) Hardness (Shore D): BPUI (46), BPUII (51), and BPUIII (65) Swelling rate (%): BPUI (4.2), BPUII (6.4), and BPUIII (5.8) Adhesion Pull-off (MPa): BPUI (2.17), BPUII (2.18), and BPUIII (0.84)	Great potential for application as coatings, and for further modification with bio-based fillers	Silva et al. [39]
Hybrid liquefied wood-based Note Orthotrichlorosilane (OTS)	König pendulum: The addition of nano silica and OTS slightly increases its hardness (LW-PU: 38 s; LW-PU-OTS: 55 s; LW-PU-0.5% Si: 40 s; LW-PU-0.5%Si-OTS: 70 s; LW-PU-1%Si: 50 s; LW-PU-1% Si-OTS: 83 s) Tensile: resistance to scratching 1–3 N Adhesion: cohesive type failure (2.16–2.27 MPa) Resistance to cold liquids: OTS-treated coatings have a good resistance to water and alcohols (48%) for 1 h duration TGA: final residue of nano silica was higher than LW-PU-OTS due to the deposition of silica in self-assembled layers Hydrophobization: contact angle of water on the OTS-treated surfaces is very stable and superhydrophobic Water droplets volume: OTS-treated coatings have a good agreement with the phenomenon of water drop volume change with time	Suitable for construction of hydrophobized liquefied wood-based polyurethane coatings	Kumar et al. [24]

coating. A significant difference in colour was measured at 48.9, indicating that a change from the dark brown colour of LW to a light-yellow colour was achieved.

In addition to that, other properties such as mass loss, hydroxyl number, and FTIR analysis were not significantly altered. Due to bleaching, it is evident that certain alterations have taken place in the mixture of liquefied wood. However, these changes did not affect the properties of the coating, and it was equivalent to the unbleached coating. Based on the literature, bleaching of liquefied biomass products was achieved by using hydrogen peroxide with potassium carbonate as a catalyst, compared with others. Hydrogen peroxide is probably the best terminal oxidant after dioxygen with respect to environmental and economic considerations [34, 45]. The bleached PU has undoubtedly found broader applications when aesthetics are the primary criterion.

References

1. Abhijit, D., Prakash, M.: A brief discussion on advances in polyurethane applications. Adv. Ind. Eng. Polym. Res. **3**(3), 93–101 (2020). https://doi.org/10.1016/j.aiepr.2020.07.002
2. Ali, E.S., Zubir, S.A.: The mechanical properties of medium density rigid polyurethane biofoam. MATEC Web Conf. **39**, 01009 (2016). https://doi.org/10.1051/matecconf/201639 01009
3. Budija, F., Tavzes, Č, Zupančič-Kralj, L., Petrič, M.: Self-crosslinking and film formation ability of liquefied black poplar. Bioresour. Technol. **100**(13), 3316–3323 (2009). https://doi.org/10.1016/j.biortech.2009.02.004
4. Chattopadhyay, D.K., Raju, K.V.S.N.: Structural engineering of polyurethane coatings for high performance applications. Prog. Polym. Sci. **32**(3), 352–418 (2007). https://doi.org/10.1016/j.progpolymsci.2006.05.003
5. Chen, S., Jiang, Y.: The acoustic property study of polyurethane foam with addition of bamboo leaves particles. Polym. Compos. **39**(4), 1370–1381 (2018). https://doi.org/10.1002/pc.24078
6. Cheumani-Yona, A.M., Budija, F., Hrastnik, D., Kutnar, A., Pavlič, M., Pori, P., Tavzes, Č, Petrič, M.: Preparation of two-component polyurethane coatings from bleached liquefied wood. BioResources **10**(2), 3347–3363 (2015). https://doi.org/10.15376/biores.10.2.3347-3363
7. Cheumani-Yona, A., Pori, P., Kričej, B., Kutnar, A., Budija, F., Tavzes, Č, Petrič, M.: Bleaching of liquefied wood for the preparation of aesthetically manageable biomaterials. J. Biobased Mater. Bioenergy **6**(5), 601–607 (2012). https://doi.org/10.1166/jbmb.2012.1254
8. Chow, M.C., Jackson, W.R., Chaffee, A.L., Marshall, M.: Thermal treatment of algae for production of biofuel. Energy Fuels **27**(4), 1926–1950 (2013). https://doi.org/10.1021/ef3 020298
9. Darabant, A., Haruthaithanasan, M., Atkla, W., Phudphong, T., Thanavat, E., Haruthaithanasan, K.: Bamboo biomass yield and feedstock characteristics of energy plantations in Thailand. Energy Procedia **59**, 134–141 (2014). https://doi.org/10.1016/j.egypro.2014.10.359
10. Deka, H., Karak, N.: Bio-based hyperbranched polyurethanes for surface coating applications. Prog. Org. Coat. **66**(3), 192–198 (2009)
11. Dimitriadis, A., Bezergianni, S.: Hydrothermal liquefaction of various biomass and waste feedstocks for biocrude production: a state of the art review. Renew. Sustain. Energy Rev. **68**, 113–125 (2017)
12. El-barbary, M.H., Shukry, N.: Polyhydric alcohol liquefaction of some lignocellulosic agricultural residues. Ind. Crops Prod. **27**(1), 33–38 (2008)
13. Fierz-David, H.E.: The liquefaction of wood and some general remarks on the liquefaction of coal. Chem. Ind. Rev. **44**, 942–944 (1925)

14. Fu, S., Ma, L., Li, W., Cheng, S.: Liquefaction of bamboo, preparation of liquefied bamboo adhesives, and properties of the adhesives. Front. For. China **1**(2), 219–224 (2006). https://doi.org/10.1007/s11461-006-0005-9
15. Gao, L.-L., Liu, Y.-H., Lei, H., Peng, H., Ruan, R.: Preparation of semirigid polyurethane foam with liquefied bamboo residues. J. Appl. Polym. Sci. **116**(3), 1694–1699 (2010). https://doi.org/10.1002/app.31556
16. Ghasemlou, M., Daver, F., Ivanova, E.P., Adhikari, B.: Polyurethanes from seed oil-based polyols: a review of synthesis, mechanical and thermal properties. Ind. Crops Prod. **142**, 111841 (2019). https://doi.org/10.1016/j.indcrop.2019.111841
17. Hassan, E.M., Shukry, N.: Polyhydric alcohol liquefaction of some lignocellulosic agricultural residues. Ind. Crops Prod. **27**(1), 33–38 (2008). https://doi.org/10.1016/j.indcrop.2007.07.004
18. Hu, S., Luo, X., Li, Y.: Polyols and polyurethanes from the liquefaction of lignocellulosic biomass. Chemsuschem **7**(1), 66–72 (2013). https://doi.org/10.1002/cssc.201300760
19. Juhaida, M.F., Paridah, M.T., Hilmi, M.M., Sarani, Z., Jalaluddin, H., Mohamad Zaki, A.R.: Liquefaction of kenaf (Hibiscus cannabinus L.) core for wood laminating adhesive. Bioresour. Technol. **101**(4), 1355–1360 (2010). https://doi.org/10.1016/j.biortech.2009.09.048
20. Kim, J.Y., Lee, H.W., Lee, S.M., Jae, J., Park, Y.K.: Overview of the recent advances in lignocellulose liquefaction for producing biofuels, bio-based materials and chemicals. Bioresour. Technol. **279**, 373–384 (2019)
21. Kim, J.-Y., Oh, S., Hwang, H., Moon, Y.-H., Choi, J.W.: Assessment of miscanthus biomass (Miscanthus sacchariflorus) for conversion and utilization of bio-oil by fluidized bed type fast pyrolysis. Energy **76**, 284–291 (2014). https://doi.org/10.1016/j.energy.2014.08.010
22. Kobayashi, M., Asano, T., Kajiyama, M., Tomita, B.: Analysis on residue formation during wood liquefaction with polyhydric alcohol. J. Wood Sci. **50**(5), 407–414 (2004). https://doi.org/10.1007/s10086-003-0596-9
23. Koul, B., Yakoob, M., Shah, M.P.: Agricultural waste management strategies for environmental sustainability. Environ. Res. **206**, 112285 (2021). https://doi.org/10.1016/j.envres.2021.112285
24. Kumar, A., Petrič, M., Kričej, B., Žigon, J., Tywoniak, J., Hajek, P., Škapin, A.S., Pavlič, M.: Liquefied-wood-based polyurethane–nanosilica hybrid coatings and hydrophobization by self-assembled monolayers of orthotrichlorosilane (OTS). ACS Sustain. Chem. & Eng. **3**(10), 2533–2541 (2015)
25. Kumar, N.: Biotechnology for Biofuels: A Sustainable Green Energy Solution. Springer Nature, Singapore (2020). https://doi.org/10.1007/978-981-15-3761-5
26. Kumar, R., Singh, S., Singh, O.V.: Bioconversion of lignocellulosic biomass: biochemical and molecular perspectives. J. Ind. Microbiol. Biotechnol. **35**(5), 377–391 (2008). https://doi.org/10.1007/s10295-008-0327-8
27. Kurimoto, Y., Doi, S., Tamura, Y.: Species effects on wood-liquefaction in polyhydric alcohols. Holzforschung **53**(6), 617–622 (1999). https://doi.org/10.1515/hf.1999.102
28. Lee, S.-H., Yoshioka, M., Shiraishi, N.: Liquefaction of corn bran (CB) in the presence of alcohols and preparation of polyurethane foam from its liquefied polyol. J. Appl. Polym. Sci. **78**(2), 319–325 (2000)
29. Li, Y., Luo, X., Hu, S.: Lignocellulosic biomass-based polyols for polyurethane applications. In: Bio-based polyols and polyurethanes. Springer briefs in molecular science, pp. 45–64. Springer, Cham (2015). https://doi.org/10.1007/978-3-319-21539-6_3
30. Lu, X., Wang, Y., Zhang, Y., Cheng, X., Yu, Y., Jin, Y.: Preparation of bio-polyols by liquefaction of hardwood residue and their application in the modification of polyurethane foams. J. Wuhan Univ. Technol.-Mater. Sci. Ed. **31**(4), 918–924 (2016). https://doi.org/10.1007/s11595-016-1468-7
31. Mudri, N.H., Abdullah, L.C., Aung, M.M., Salleh, M.Z., Awang Biak, D.R., Rayung, M.: Comparative study of aromatic and cycloaliphatic isocyanate effects on physico-chemical properties of bio-based polyurethane acrylate coatings. Polymers **12**(7), 1494 (2020)
32. Papaj, E.A., Mills, D.J., Jamali, S.S.: Effect of hardener variation on protective properties of polyurethane coating. Prog. Org. Coat. **77**(12), 2086–2090 (2014)

33. Patil, A.M., Jirimali, H.D., Gite, V.V., Jagtap, R.N.: Synthesis and performance of bio-based hyperbranched polyol in polyurethane coatings. Prog. Org. Coat. **149**, 105895 (2020)
34. Pędziwiatr, P., Mikołajczyk, F., Zawadzki, D., Mikołajczyk, K., Bedka, A.: Decomposition of hydrogen peroxide - kinetics and review of chosen catalysts. Acta Innov. **26**, 45–52 (2018). https://doi.org/10.32933/actainnovations.26.5
35. Petrič, M., Sernek, M.: Bonding and surface finishing of wood with liquefied wood. Pro Ligno **11**(4), 2015 (2015)
36. Sadiku, N., Oluyege, A., Sadiku, I.: Calorific and fuel value index of bamboo. Lignocellulose **5**(1), 34–49 (2016)
37. Sharma, R., Wahono, J., Baral, H.: Bamboo as an alternative bioenergy crop and powerful ally for land restoration in Indonesia. Sustainability **10**(12), 4367 (2018). https://doi.org/10.3390/su10124367
38. Shen, S., Nges, I.A., Yun, J., Liu, J.: Pre-treatments for enhanced biochemical methane potential of bamboo waste. Chem. Eng. J. **240**, 253–259 (2014). https://doi.org/10.1016/j.cej.2013.11.075
39. Silva, T.A., Marques, A.C., Dos Santos, R.G., Shakoor, R.A., Taryba, M., Montemor, M.F.: Development of biopolyurethane coatings from biomass-derived alkylphenol polyols—a green alternative. Polymers **15**(11), 2561–2561 (2023). https://doi.org/10.3390/polym15112561
40. Soares, B., Gama, N., Freire, C., Barros-Timmons, A., Brandão, I., Silva, R., Pascoal Neto, C., Ferreira, A.: Ecopolyol production from industrial cork powder via acid liquefaction using polyhydric alcohols. ACS Sustain. Chem. & Eng. **2**(4), 846–854 (2014). https://doi.org/10.1021/sc400488c
41. Song, Y., Lee, Y.G., Ahn, Y.S., Nguyen, D.-T., Bae, H.-J.: Utilization of bamboo as biorefinery feedstock: co-production of xylo-oligosaccharide with succinic acid and lactic acid. Bioresour. Technol. **372**, 128694 (2023). https://doi.org/10.1016/j.biortech.2023.128694
42. Sun, F., Li, X., Zhang, Q.: Technologies of liquefaction of bamboo and preparation of adhesive. Front. For. China **3**(4), 493–498 (2008). https://doi.org/10.1007/s11461-008-0076-x
43. Sun, Y., Gao, B., Yao, Y., Fang, J., Zhang, M., Zhou, Y., Chen, H., Yang, L.: Effects of feedstock type, production method, and pyrolysis temperature on biochar and hydrochar properties. Chem. Eng. J.m. Eng. J. **240**, 574–578 (2014). https://doi.org/10.1016/j.cej.2013.10.081
44. Szycher, M.: Szycher's Handbook of Polyurethanes. CRC Press (2013). https://doi.org/10.1201/b12343
45. Tutus, A.: Bleaching of rice straw pulps with hydrogen peroxide. Pak. J. Biol. Sci. **7**(8), 1327–1329 (2004). https://doi.org/10.3923/pjbs.2004.1327.1329
46. Velayutham, T.S., Majid, W.H.A., Ahmad, A.B., Kang, G.Y., Gan, S.N.: Synthesis and characterization of polyurethane coatings derived from polyols synthesized with glycerol, phthalic anhydride and oleic acid. Prog. Org. Coat. **66**(4), 367–371 (2009). https://doi.org/10.1016/j.porgcoat.2009.08.013
47. Wang, Y.-N., Ko, C.-H., Lee, C.-Y., Tsai, H.-P., Chen, W.-H., Hwang, W.-S., Tsai, M.-J., & Chang, F.-C. (2011). Evaluation of bamboo as a feedstock for bioethanols in Taiwan. In: World Renewable Energy Congress – Sweden, 8–13 May 2011, Linköping, Sweden. https://doi.org/10.3384/ecp11057171
48. Xie, J., Qi, J., Hse, C.-Y., Shupe, T.F.: Effect of lignin derivatives in the bio-polyols from microwave liquefied bamboo on the properties of polyurethane foams. BioResources **9**(1), 578–588 (2014)
49. Xu, L., Liao, B., Nian, F.W., Huang, J.H., Guo, D.D., Deng, M.J.: Liquefaction of bamboo powder in polyols and its application in preparation of semi-rigid polyurethane. Chem. Indus. For. Prod. **33**(6), 48–54 (2013)
50. Yan, Y., Hu, M., Wang, Z.: Kinetic study on the liquefaction of cornstalk in polyhydric alcohols. Ind. Crops Prod. **32**(3), 349–352 (2010). https://doi.org/10.1016/j.indcrop.2010.05.015
51. Yang, R., Wang, B., Xu, L., Zhao, C., Zhang, X., Li, J.: Synthesis and characterization of rigid polyurethane foam with dimer fatty acid-based polyols. Polym. Bull. **76**, 3753–3768 (2019)
52. Ye, L., Zhang, J., Zhao, J., Tu, S.: Liquefaction of bamboo shoot shell for the production of polyols. Bioresour. Technol. **153**, 147–153 (2014). https://doi.org/10.1016/j.biortech.2013.11.070

53. Zhang, J.P., Du, M.H., Hu, L.S.: Liquefaction of bamboo polyhydric alcohols and its application in polyurethane foam materials. Adv. Mater. Res. **781–784**, 632–635 (2013). https://doi.org/10.4028/www.scientific.net/amr.781-784.632
54. Zhang, T., Zhou, Y., Liu, D., Petrus, L.: Qualitative analysis of products formed during the acid catalyzed liquefaction of bagasse in ethylene glycol. Bioresour. Technol. **98**(7), 1454–1459 (2007). https://doi.org/10.1016/j.biortech.2006.03.029

Environmental and Economic Impacts of Green Panels

Nur Izzah Nabilah Haris, Syeed SaifulAzry Osman Al Edrus, Norfaryanti Kamaruddin, Nurliyana Abdul Raof, A. W. Noraida, and Mohd Hafizz Wondi

Abstract This chapter embarks on the multifaceted nature of the environmental impact of green lignocellulosic-based panels. It begins with the critical selection of eco-friendly materials and exploring bio-based binders like tannin-based bio resins and modified starch. Technological advancements in greener pretreatment methods, such as supercritical fluids and ionic liquids, add depth to the sustainable manufacturing narrative. The manufacturing phase, focusing on closed molding processes, emerges as a key player in reducing environmental impact. Despite recycling challenges posed by binders, proactive measures like eco-friendly binders and binderless technology are promising solutions for efficient end-of-life recycling. Anchored in life cycle assessment principles, particularly the use of life cycle assessment to comprehensively evaluate environmental footprints, this exploration lays a foundation for informed decision-making, fostering a holistic understanding of green panel sustainability. Another important factor to take into account is the economic impact of these sustainable activities. It is emphasized throughout the chapter that

N. I. N. Haris (✉) · S. S. Osman Al Edrus
Institute of Ecosystem Science Borneo, Universiti Putra Malaysia Bintulu Campus, Bintulu, Malaysia
e-mail: izzahnabilah@upm.edu.my

N. I. N. Haris
Department of Chemical and Environmental Engineering, Faculty of Engineering, Universiti Putra Malaysia, Serdang, Malaysia

N. Kamaruddin
Institute of Tropical Forestry and Forest Products (INTROP), Universiti Putra Malaysia, Serdang, Malaysia

N. Abdul Raof
Chemical Engineering Department, Faculty of Engineering, Universiti Teknologi PETRONAS, Perak, Malaysia

A. W. Noraida
Centre of Excellence for Research Innovation (CERI), MRANTI Technology Park, Kuala Lumpur, Malaysia

M. H. Wondi
Faculty of Plantation and Agrotechnology, Universiti Teknologi MARA Sarawak Branch Mukah Campus, Mukah, Malaysia

S. S. Osman Al Edrus et al. (eds.), *Green Lignocellulosic-Based Panels*, Engineering Materials, https://doi.org/10.1007/978-981-96-3404-0_10

implementing eco-friendly materials and technology can have a favorable impact on industry economics in addition to supporting ecological sustainability. Together with lower regulatory compliance costs, increased brand recognition, and the opening up of new markets for environmentally friendly products, these factors will eventually promote a more robust and sustainable business climate.

Keywords Environmental impact · Life cycle assessment · Environmental footprint · Green panels

1 Introduction

In an era where sustainability has taken the center stage, green lignocellulosic-based panels are becoming the spotlight, capturing the attention of researchers and industries. Green panels, regarded as environmentally friendly, have emerged as a promising alternative to conventional materials in various applications such as construction, furniture, automotive, packaging, and insulation. At first glance, looking only at their raw materials might lead one to perceive green lignocellulosic panels as superior to their conventional counterparts in terms of environmental impact. However, the utilization of renewable resources often comes with its own set of trade-offs, such as potential reduction in performance, increased binder requirements, and additional manufacturing processes. This necessitates a closer examination and a detailed understanding of the environmental implications associated with the entire life cycle of these panels—from production and use to disposal. This chapter embarks on the multifaceted nature of the environmental impact of green lignocellulosic-based panels. By delving into the life cycle of green lignocellulosic-based panels, we aim to provide a comprehensive and tangible indication of their environmental footprint.

Before exploring the environmental impacts, let's establish a clearer understanding of what "green" signifies in the context of green panels. What key factors contribute to the label of "green"? While a standardized definition for green lignocellulosic panels might not exist, at least as of the current writing of this book, it is generally acknowledged that these panels are composed of at least one renewable and sustainable raw material [31]. For instance, [28] innovatively engineered green composite panels using *Triumfetta cordifolia* fibers and utilized tannin from *Aningeria altissima* bark as bio resins. Similarly, [17] developed a green hybrid composite with wood particles and glass fiber core and a woven jute fabric skin. The presence of at least one component sourced from renewable resources designates the panel as "green". Beyond raw materials, various factors contribute to labeling panels as "green", such as efficient and eco-friendly manufacturing processes, sustainable sourcing of raw material, biodegradability, recyclability, and an overall low environmental footprint. In the absence of a tangible indication of a panel's "greenness," considering these factors underscores the true ecological consciousness of the panels.

Several studies have focused on these aspects of green panels. A groundbreaking work by [25] examined an environmentally benign pretreatment technique for oil palm fronds, utilizing ionic liquid treatment for green composite board manufacturing. Unlike conventional lignocellulosic materials, pretreatments such as acid treatment, alkali treatment, organosolv process, and ozonolysis that involve hazardous and toxic chemicals, employing green treatment methods for board manufacturing contributes to the production of genuinely green panels. In another insightful study, [44] investigated the environmental impact of green wooden boards bonded with lignin-based binder using life cycle assessment (LCA). Their findings highlighted hydrogen peroxide (H_2O_2) production as having the most significant environmental impact. However, a comparative analysis with conventional petroleum-based bonded panels revealed that the latter exhibited even more adverse environmental impacts. These findings underscore the importance of considering the entire life cycle and various factors in assessing the true "green" nature of lignocellulosic panels.

In this chapter, our goal is to explore the environmental impact and economic impact of green lignocellulosic-based panels, shedding light on the intricacies that define their ecological and financial footprint. As the importance of sustainability grows in our current era, we aim to provide a thorough understanding of what makes these panels genuinely "green". The chapter serves as a foundation for a deeper examination, touching on aspects such as the use of LCA for environmental impact and economic assessment, sustainable sourcing of raw materials, eco-friendly manufacturing technologies, and recycling challenges and opportunities. By discussing the factors that contribute to the "green" designation and referencing significant studies, we lay the groundwork for a detailed look into the many environmental and economic considerations these panels face throughout their life cycle.

2 Life Cycle Assessment of Green Panels

Determining the environmental friendliness of a product or material hinges on a scientifically validated approach that rigorously assesses its overall impact—from the extraction of raw materials to production, transportation, and disposal. Life Cycle Assessment (LCA) emerges as a robust methodology explicitly designed for this purpose, offering a comprehensive examination of a product's environmental footprint from cradle to grave. This holistic perspective delves into the intricacies of each life cycle stage, facilitating an understanding of both the environmental burdens and benefits associated with the product. The International Organization for Standardization (ISO) has laid the groundwork for LCA through ISO 14040, delineating the concept and framework, and ISO 14044, outlining the procedure and requirements [20, 21]. The sequential phases of LCA, encompassing goal and scope definition, inventory analysis, impact assessment, and interpretation, are systematically coordinated to provide a comprehensive and scientifically sound evaluation [18].

2.1 Goal and Scope Definition

Goal and scope definition is the first step of conducting LCA, setting the foundation for a holistic environmental evaluation. This crucial step involves outlining the assessment's purpose, defining the functional unit, and establishing the system boundaries. In the context of green lignocellulosic panels, the goal is distinctly defined: to meticulously evaluate the environmental performance across the entire life cycle. In situations where disposal information is either unavailable or disregarded, researchers often opt for a cradle-to-gate approach, concentrating solely on the manufacturing facet of the product. An illustrative example is found in [7] exploration of the environmental impact of eucalyptus bark panels manufactured for insulation applications, emphasizing the manufacturing stage. Another notable study by [33] delves into a comparative LCA for neat PLA, and PLA containing kenaf or pine needle fibers bio-pellets intended for the production of composite panels.

The determination of the functional unit is equally pivotal at this stage, quantifying the amount of the product under analysis and forming the basis for discerning input and output flows in processes. Researchers like [13] opted for a functional unit of 1 m^2 of lath made from olive pomace composite in their LCA assessment, while [41] selected 1 kg of wood–plastic composite as their reference. In the case of lignocellulosic panels designed for specific applications, such as thermal insulation, the functional unit aligns with the capacity of thermal resistance. For instance, [7] set their functional unit for eucalyptus bark insulation panels as the mass (kg) of the panel that provides 1 m^2K/W of thermal resistance.

Afterwards, it is necessary to set the system boundaries of the assessment, from the extraction of lignocellulosic fibers and the manufacturing process to transportation, use, and disposal. The system boundaries for several LCA studies of lignocellulosic panels are depicted in Fig. 1, elucidating the breadth and depth of the assessments. For example, [7] set a cradle-to-gate system boundary, encapsulating the extraction of bark involving forest management and the wood chip process, the subsequent transport of biomass to the production facility, and the manufacturing stages culminating at the factory gate. It is noteworthy that certain phases, such as panel installation, maintenance, and end-of-life disposal, have been deliberately excluded from consideration. This deliberate omission is attributed to the innovative nature of the insulation material under investigation, resulting in a lack of available data for these specific life cycle stages.

2.2 Life Cycle Inventory

The Life Cycle Inventory (LCI) stage of LCA involves identifying the input and output flows of the system, providing a detailed inventory of material and energy flows [30]. The quantification is made based on the functional unit specified in the

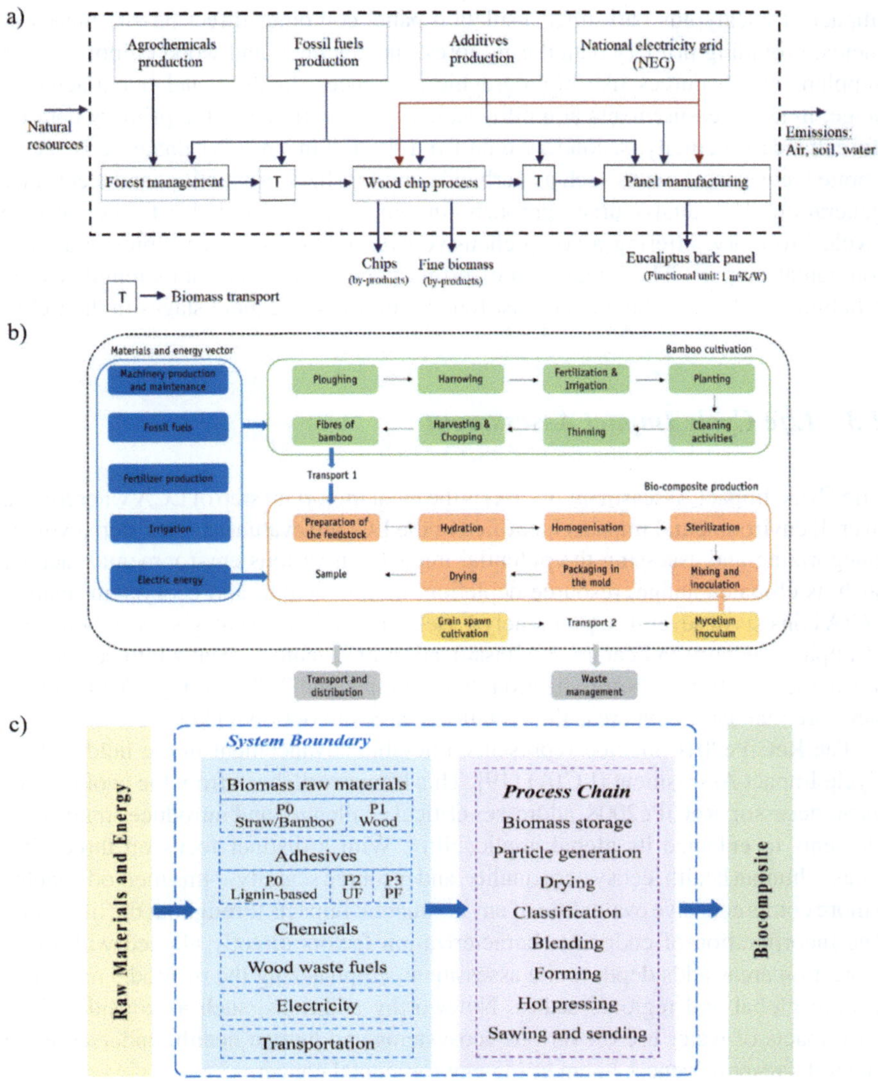

Fig. 1 LCA system boundaries of **a** eucalyptus bark fiber insulation panel, **b** bamboo–mycelium composite, and **c** rice straw–bamboo composite. Reproduced with permission [6, 7, 36] [Creative Commons Attribution (CC BY) license]

previous step. The resource consumption, energy use, emissions, and waste generation of each stage of life of green lignocellulosic panels should be evaluated. Data collection for a comprehensive and detailed inventory creation typically requires a dual-source approach: primary and secondary. Primary data collection involves on-site measurement whereas secondary data collection involves the use of data from databases. For example, the strategy employed by [7] for assessing the environmental impact of eucalyptus bark fiber insulation panel combines direct on-site measurements, providing primary data during forest management and wood chipping, with supplementary sources like bibliographic references. In the panel manufacturing stage, in situ measurements at a pilot-scale plant contribute to the primary dataset. To enhance specificity, secondary data from the Ecoinvent v2.1 database is incorporated, covering aspects such as herbicide and fertilizer production, and electricity generation. This dual-source approach aims to create a detailed yet efficient Life Cycle Inventory, offering a comprehensive overview of resource inputs and environmental outputs. The integration of primary and secondary data strengthens the reliability of the LCI, laying a robust foundation for subsequent stages in the LCA.

2.3 Life Cycle Impact Assessment

Life Cycle Impact Assessment (LCIA) is the most important step of LCA whereby the overall environmental impacts identified in the LCI are evaluated. This step involves categorizing and assessing the potential impacts on various environmental factors, such as climate change, resource depletion, human health, and ecosystem quality. LCIA consists of several steps, namely 1. Selection of impact categories, 2. Selection of impact category indicators, 3. Classification of inventory results into categories, 4. Characterization, 5. Normalization, 6. Grouping, and 7. Weighting. The first four steps are mandatory, whereas the remaining three are optional [11].

The ReCiPe2016 method represents a notable advancement in the field of Life Cycle Impact Assessment (LCIA) [19]. This framework, born from the evolution of its predecessor, ReCiPe2008, addresses critical challenges and introduces innovative elements to enhance its global applicability. With a distinct focus on three vital areas—human health, ecosystem quality, and resource scarcity—this method enables a more comprehensive evaluation of environmental impacts throughout the life cycle. The incorporation of endpoint characterization factors directly aligned with these protection areas adds depth to the assessment, emphasizing the method's relevance at both global and regional scales. Noteworthy additions, such as considerations for impacts of water use on various ecosystems and human health, underscore the method's responsiveness to emerging environmental concerns.

A work by [13] utilized the ReCiPe2016 method to conduct the LCIA of olive pomace biocomposites with two different matrices, polypropylene (PP) and polyethylene (PE). The twin-screw compounding process, heavily influenced by polymer production, stands out as a significant contributor to most midpoint impact categories in both scenarios. Figure 2 highlights the comparison of contributions of each process to the environmental impact for both composite materials. Results highlight human health as the most affected area, constituting 89% of the total impact in both scenarios, with energy, carbon dioxide, and sulfur dioxide from manufacturing as key contributors. Ecosystems and resource availability contribute less at 6% and 4.5%, respectively. Furthermore, the biocomposites' prolonged lifespan allows for carbon storage, mitigating climate change by delaying carbon emissions. This case

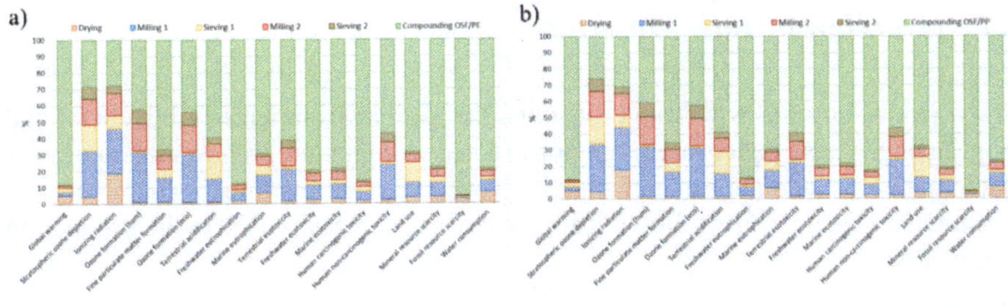

Fig. 2 Environmental impact from each process of **a** olive pomace-PE and **b** olive pomace-PP composites [13] [Creative Commons Attribution (CC BY) license]

study, employing the ReCiPe2016 method, underscores the potential of olive pomace-based composites as eco-friendly alternatives in the building sector, prompting future exploration with bio-sourced polymer matrices to further minimize environmental impact.

2.4 *Interpretation of Results*

The interpretation phase in the LCA is the final step wherein results from prior stages are comprehensively synthesized and evaluated. This phase seeks to gain meaningful insights from the intricate environmental data amassed across the life cycle of a given product or system. Its primary objective is to contextualize findings, identify noteworthy observations, and draw informed conclusions about environmental performance [18]. A proficient interpretation informs decision-makers and offers guidance for future actions, with a focus on elevating sustainability and mitigating environmental impacts of a product.

3 Sustainable Sourcing of Raw Materials

Green panels, positioned as environmentally friendly alternatives to traditional wood-based panels, represent a crucial stride in mitigating the environmental impact of the construction and furniture sectors. While the sustainability of these panels is intricately linked to production methods and the use of eco-friendly materials, a key determinant is the sourcing of raw materials. Sustainable raw material sourcing not only minimizes the adverse environmental effects of resource extraction but also ensures the preservation of the natural equilibrium. Additionally, delving into the economic impact, efficient raw material procurement contributes to cost-effectiveness, thereby enhancing the overall economic viability of green panel production.

3.1 Overview of Raw Materials

Green panels cover a wide range of items that are environmentally beneficial and long-lasting. These panels are often constructed from natural, renewable, and recyclable materials, and their manufacturing procedures use fewer toxic chemicals and consume less energy. The following are the key raw materials used in the production of green panels:

i. **Wood**: The most prevalent raw material for panel manufacture is wood, and its sustainability is critical for producing green panels. Sustainable forestry practices such as selective harvesting, reforestation, and conservation are critical for ensuring long-term supply of wood resources while minimizing environmental effect (Rainforest [37]).
ii. **Recycled wood**: Recycled wood is a useful resource that can reduce the demand for virgin wood extraction greatly. Manufacturers of green panels are increasingly using reclaimed wood into their products, diverting wood waste from landfills, and encouraging a circular economy [40].
iii. **Crops and agricultural residues**: Agricultural wastes such as straw, coconut husk, and bamboo have the potential to be viable alternatives to wood-based fibers. These materials are plentiful, renewable, and frequently possess special features that can improve the performance of green panels (Rainforest [37]).
iv. **Natural binders**: As alternatives to formaldehyde-based adhesives, natural binders such as soy protein resin, starch-based adhesives, and lignin-based resins are increasingly being employed in green panel manufacturing. These binders have better environmental performance and are less toxic (Rainforest [37]).

3.2 Sustainable Forestry Practices

Sustainable forestry practices are critical for preserving the long-term viability of wood supplies and reducing the environmental effect of raw material extraction for green panel manufacture. These practices strive to sustain forest health and production while balancing the forestry industry's economic needs with the preservation of ecosystem health and biodiversity. Some of the most important features of sustainable forestry practices are as follows (Rainforest [37]):

i. **Planning and management**: To be sustainable, forest resources must be carefully planned and managed, considering the ecological, social, and economic values of forests. Setting defined objectives, monitoring forest conditions, and applying adaptive management measures to respond to changing circumstances are all part of this process.
ii. **Harvesting and regeneration**: Sustainable forestry guarantees that wood harvesting does not exceed forests' natural regeneration capacity, and that harvesting practices minimize forest ecosystem damage. This includes using

proper silvicultural techniques such as thinning, pruning, and coppicing, as well as ensuring natural or artificial regeneration of harvested areas.

iii. **Conservation and protection**: Sustainable forestry safeguards forests' biological integrity and diversity, as well as the habitats and species of flora and wildlife that rely on them. This includes identifying and protecting high conservation value forests such as old-growth forests, riparian zones, and wildlife corridors, as well as preventing and controlling fire, pests, diseases, and invasive species threats.

iv. **Certification and verification**: Independent third-party certification systems analyze forestry operations' compliance with requirements for environmental, social, and economic sustainability. These approaches assure consumers and stakeholders that wood products originating from certified forests are produced responsibly.

3.3 Certification Systems

Certification methods are crucial in promoting sustainable forestry practices and ensuring that raw materials used in green panel manufacture are responsibly obtained. These systems allow independent verification that forestry operations follow environmental, social, and economic sustainability standards. Some of the most recognized green panel certification systems are [39]

i. **Forest Stewardship Council (FSC)**: The Forest Stewardship Council (FSC) is a global non-profit organization that promotes responsible forest management through its Principles and Criteria, which address environmental, social, and economic elements of forestry. FSC certifies forests and forest products and maintains a chain of custody systems to track the origin and movement of certified wood products.

ii. **Program for the Endorsement of Forest Certification (PEFC)**: PEFC is a non-profit international organization that supports national and regional forest certification systems that meet its Sustainability Benchmarks, which are based on internationally agreed-upon criteria for sustainable forest management. PEFC certifies forests and forest products as well as maintains a chain of custody systems that tracks the origin and flow of certified wood products.

iii. **Malaysian Timber Certification Scheme (MTCS)**: The Malaysian Timber Certification Scheme (MTCS) is a national forest certification system that is supported by PEFC and applies to both natural and cultivated forests in Malaysia. The Malaysian Criteria and Indicators for Forest Management Certification, which are matched with the ITTO Criteria and Indicators for Sustainable Management of Tropical Forests, form the foundation of MTCS. MTCS certifies forests and forest products, as well as maintains a chain of custody systems that tracks the origin and flow of certified wood products.

3.4 Responsible Raw Material Procurement

Responsible raw material procurement is also crucial in the context of mitigating climate change impacts. The study by [27] highlights the role of sustainable forest management in sequestering carbon and promoting resilience to climate-related challenges. By ensuring responsible harvesting, reforestation, and biodiversity conservation, these practices contribute to the net removal of carbon dioxide from the atmosphere. Green panel manufacturers, relying on raw materials sourced from sustainably managed forests, align themselves with this carbon sequestration objective. This alignment not only reinforces their commitment to environmental sustainability but also positions them as active participants in climate change mitigation. From an economic perspective, research by [35] underscores the financial benefits of sustainable sourcing. Efficient raw material procurement reduces supply chain risks, ensuring a stable and reliable influx of materials. Furthermore, embracing sustainable practices allows manufacturers to tap into the growing market demand for eco-friendly products, fostering economic growth within the industry.

Green panel manufacturing emerges as a promising solution to alleviate the environmental impact of the construction and furniture industries, but its success hinges on responsible raw material procurement. Sustainable forestry practices are paramount, ensuring the long-term availability and quality of wood supplies while simultaneously reducing the environmental impact of raw material exploitation. Certification systems play a crucial role, providing a dependable and transparent method to verify the sustainability of forestry operations and wood products. By adhering to such certifications, green panel manufacturers not only demonstrate their commitment to environmental responsibility but also contribute to meeting the growing market demand for sustainably sourced materials, thereby reinforcing the industry's dedication to eco-friendly practices.

4 Eco-Friendly and Economically Effective Manufacturing Technologies

As technological improvements promote innovation in the production of green lignocellulosic-based panels, there is a growing need to address the environmental impact in the manufacturing process. The pressing need for a sustainable manufacturing process and the adoption of green processing technologies has fueled research efforts targeted at reducing environmental impact while maintaining or improving the functionality of manufactured panels. This section aims to investigate the recent advancements and cutting-edge research in sustainable manufacturing and green processing technologies in the development of green panels. The economic dimension of this exploration lies in the potential for cost-effectiveness and market competitiveness. Sustainable manufacturing practices not only align with environmental goals but also present opportunities for operational efficiency and resource optimization, which can result in cost savings. Apart from conventional manufacturing processes, the most often used fabrication techniques for green composites include thermoforming, pultrusion, injection, compression, extrusion, and resin transfer molding (RTM) which offer diverse options for manufacturers to choose methods that suit both environmental and economic objectives [14]. All these techniques can be broadly classified as open or closed molding, depending on the application procedure of both the fiber material and the binder's matrix [1]. For instance, closed molding processes, by reducing material waste and energy consumption, contribute to cost-effectiveness in the long run. Understanding the economic advantages of different techniques can guide manufacturers in selecting the most viable and sustainable options for their specific applications. The fabrication of green composites via compression molding and thermoforming can be divided into the following four basic steps: material selection, material preprocessing, semi-finished product manufacturing, and fabrication of green composite.

4.1 Material Selection

Choosing the right materials for green lignocellulosic-based panels is a crucial step in the manufacturing process with profound implications for both environmental sustainability and economic impact. One of the most important attributes of the green panels is their biodegradability, which enables them to penetrate as an environmental solution into the global market. A great deal of study has been done to determine which natural fibers are best for use as reinforcement and to improve the qualities of the produced green panels. Lignocellulosic-based panels are generally produced from adding wood or non-wood lignocellulosic materials with an adhesive or binder, and thermally pressing them together to form a flat composite panel [34]. The economic impact is intricately tied to the choice of binders and adhesives in the production process. The binders or adhesive are essential in holding the fibers together, creating a solid and stable board. Although they serve different purposes, adhesive and binders are commonly used interchangeably in the literature. The choice of binders and adhesives can significantly impact the properties and environmental friendliness of the green panels. More research is being conducted to reduce or eliminate the use of synthetic binders and synthetic adhesive in the production process in order to make it more environmentally friendly [9]. Most synthetic binders used in fiberboard production are formaldehyde-based. It is derived from non-renewable resources and is classified as carcinogenic to humans [12, 42]. The economic implications are noteworthy, as the shift toward natural, renewable resources not only aligns with environmental goals but also responds to market demands for sustainable and eco-friendly products. Consequently, the exploration of alternative binders and adhesives not only contributes to the environmental friendliness of green panels but also holds the potential to reshape economic dynamics within the manufacturing industry.

The pursuit of a completely biodegradable composition in green panels not only aligns with environmental objectives but also carries significant implications from an economic perspective. The most widely used technique for achieving a completely biodegradable composition in green panels is probably the use of bio-based binders, such as starch- or protein-based binders obtained from sources like milk or gelatine [34]. For instance, [8] used modified starch and a little amount of urea formaldehyde as a binder to create particleboard panels from Eastern red cedar (*Juniperus virginiana L.*). Their findings suggested the potential of modified starch as a green binder in particleboard panels without having a significant negative impact on their qualities. The development of tannin-based bio resins for use in particleboard and fiberboard has been mentioned in a few of the research [28]. Because tannin-based bio resins are more reactive than phenols, they provide bonding capabilities that are comparable to synthetic resins but with less environmental impact. Another study from [3] found that eco-friendly plywood samples made from four distinct wood species (beech, birch, hornbeam, and poplar) and bonded with LDPE film had satisfactory physical–mechanical qualities. The discovery of using citric acid (CA) as a green modifying agent and adhesive for wood veneer panels was quite groundbreaking and thrilling [10]. The environmentally modified wood with CA exhibited improvement in modulus of elasticity, compression strength, dimensional stability, and biological resistance. Modifying wood with CA, a green renewable chemical present in fruits and vegetables, is predicted to improve product performance while lowering the ultimate environmental imprint [24]. In essence, the exploration of bio-based binders and green modifying agents not only contributes to the environmental sustainability of green panels but also holds the potential to drive economic efficiency and competitiveness within the industry.

In a different study, the potential of natural kenaf fiber-reinforced concrete containing LC^3 binder as a green cementitious composite was investigated by [2]. The idea was to replace cement with supplementary material as additive to reduce the GHG emission and carbon footprint. These formulations have shown great potential as green composite panels due to its lower energy consumption, reduced environmental impacts, and improved mechanical and durability properties. Recent research has shown that binderless fiberboards can offer a higher density and improved physio-mechanical properties compared to traditional particleboards produced with binders [29]. The use of binderless fiberboards promotes forest preservation, while the absence of formaldehyde emissions makes these panels particularly appropriate for indoor uses such as wall partitions, fake ceilings, and furniture [42]. In summary, bio-based adhesives and binders for green panels offer potential alternatives to standard synthetic adhesives, with enhanced environmental impacts and performance features. While green composites made from natural sources have the potential to reduce environmental impact compared to traditional composites, it is important to carefully consider and address the environmental issues associated with their production to maximize their sustainability and minimize negative impacts on the environment.

4.2 Material Preprocessing

In green panel manufacturing, the material preprocessing steps usually consist of biomass fiber preparation, drying, and pretreatment. Because pretreatment is such an important and challenging undertaking, this chapter will concentrate on its environmental impact. Pretreatment must be done to improve lignocellulosic biomass digestibility by weakening the complex network of chemical bonds between cellulose, hemicellulose, and lignin [38]. A variety of traditional pretreatment methods, including physical, physicochemical, chemical, and combined treatments, are available. The majority of these processes necessitate high temperatures and pressures, as well as highly concentrated acids and bases [25, 38]. Greener pretreatment technologies that use "greener" solvents including supercritical fluid, ionic liquid, and deep eutectic solvent are becoming more popular. They provide milder reaction conditions, employ less harmful chemicals, consume significantly less energy, and generate significantly less waste [38].

The use of supercritical carbon dioxide in the pretreatment of industrial hemp bast fibers was reported to increase fiber fineness and decrease moisture sensitivity while retaining initial temperature resistance [15]. High temperatures and pressures have been used for hot water and supercritical fluid pretreatments, but no harmful chemicals are needed for these procedures [43]. Ionic liquids (IL) have shown promise in green processing technology for panel production too due to their unique and green properties. IL is the potential alternative to environmentally harmful organic solvents due to its noncorrosive and non-volatile nature, having excellent dissolution power under relatively mild conditions and high thermal stability [26]. The pretreatment of ILs [bmim][Cl] (1-butyl-3-methylimidazolium chloride) and [emim][dep] (1-ethyl-3-methylimidazolium diethyl phosphate) on green composite board fabricated from oil palm frond lignocellulosic residue and thermoplastic starch binder was studied by [25]. Thermal stability was reported to be increased, with superior strength and modulus compared to untreated composite boards.

4.3 Semi-finished Product Manufacturing

Semi-finished products of green composite can be in the form of granules, sheets, prepregs, or laminates [22]. The two types of technologies used in the production of semi-finished products are fiber mat technology (impregnation with thermoplastic or thermoset resins to form sheets or prepregs) and compounding (melt mixing and extrusion of semi-finished products). Both technologies contribute to the development of environmentally friendly and cost-effective semi-finished products for fiber-reinforced bio-based composites, which are designed to have a smaller carbon footprint and conserve resources.

4.4 *Fabrication of Green Composite*

Semi-finished products, such as sheets or prepregs, are utilized to create finished panels in this step. As mentioned previously, the fabrication of green composite can be divided into two general methods, open and closed molding. The distinction between these two methods is based on how the resin is cured. Open molding occurs when the resin is exposed to the atmosphere during the curing process. It involves the placement of the reinforcement mats and the resin infusion in an open mold. Closed molding occurs when the resin is not exposed to the atmosphere during cure. Closed molding begins with the placement of dry reinforcements in the base mold, followed by the closure of the mold and the introduction of resin into the closed cavity via a pressure pump or vacuum. The mold is opened, and the part is removed once the laminate has been cured [5]. Closed molding processes generally offer lower emissions, reduced waste, and higher production efficiency compared to open molding processes, making them a more environmentally friendly option for green composites.

5 Recycling Challenges and Opportunities

The inherent sustainability of green lignocellulosic panels lies in their composition, typically derived from renewable sources such as wood fibers or agricultural residues. However, recycling these panels poses unique challenges. Unlike conventional materials, the recycling infrastructure for lignocellulosic panels remains less established on an industrial scale, contributing to limited global reuse and recycling activities for green panels [45]. Nevertheless, there is an increasing trend of recycling activity in the European Union, driven by systematic legislative policies [45]. This burgeoning trend aligns with the economic perspective, emphasizing the growing recognition of circular economy principles in waste management. The exploration of recycling considerations for green panels becomes integral to fostering a circular economy, focusing on optimized material use and waste reduction.

Composite panels can be recycled via several methods such as mechanical, thermos-hydrolytic, and chemical (or combinations thereof) [23]. However, recycling green lignocellulosic panels at the end of life possess some challenges. Notably, the presence of various adhesives and binders in the panels can complicate the recycling process. Fu et al. [16] attempted to recycle urea–formaldehyde resin-bonded particleboards. To recover the fibers, they used a thermo-hydrolytic process (hydrolysis using steam and pressure) and successfully removed 65% of the resin, facilitating the recovery of fibers. New particleboards made entirely from recycled materials demonstrated properties comparable to those made with fresh particles. Despite this success, the energy and resource consumption involved in resin removal necessitate consideration, potentially posing a more impactful ecological footprint compared to manufacturing particleboards from fresh particles. Additionally, achieving complete resin removal remains challenging, signifying the complexity of recycling lignocellulosic panels due the presence of binders and adhesives, and possibly economically inefficient. Another significant challenge in recycling panels is the separation of these panels from other attached or unattached components during the dismantling phase [7]. For example, in the automotive industry, end-of-life cars are often sent to disposal sites and shredded, making separation from metallic materials and glass a sophisticated process. Despite the feasibility of such separation, the polymeric mixture often requires an additional separation process prior to recycling [45].

Despite the challenges, efforts to address the issues associated with recycling green panels are underway. Researchers and industry stakeholders are actively engaged in exploring innovative solutions to enhance the efficiency and feasibility of recycling processes for lignocellulosic panels. Recently, the development of recycling technologies specific for green panels has gained traction. Three alternatives for end-of-life panels, apart from disposal in landfills, include physical recycling, (bio)chemical recycling, and waste to energy conversion [46]. Physical recycling usually involves disintegrating the components of panels to expose the fiber. This method avoids chemical alteration of the material, maintaining its desired properties. (Bio)chemical recycling, on the other hand, breaks down the matrix into its original raw materials. Pyrolysis and enzymolysis are examples of chemical treatment to recycle lignocellulosic panels.

From a long-term economic standpoint, investing in the development of efficient recycling technologies for green panels not only aligns with environmental goals but also presents opportunities for job creation and the emergence of a secondary market for recycled panel materials. Besserer et al. [4] highlights the economic advantages of recycling technologies, emphasizing how the development of efficient systems can contribute to job creation and the establishment of a secondary market. The emergence of a secondary market for recycled panel materials aligns with the findings of [32], who discussed the economic feasibility of recycling wood-based materials, suggesting that recycling initiatives can lead to cost savings and economic opportunities. As recycling infrastructure matures, it may unlock economic incentives for manufacturers which explores the economic implications of recycling technologies in the context of sustainability.

Moreover, the emergence of the circular economy concept has been a strong impetus to develop green panels with recycling in mind. Proactive measures involve the use of eco-friendly binders and adhesives that facilitate easier separation during recycling. Besides, binderless panels also stand out as an interesting solution in the context of sustainable panels. Binderless technology is the production of boards without the use of neither synthetic nor green binders. Rather than relying on external binders to form a solid structure, the lignocellulosic raw material within the boards becomes the adhesive itself. Integrating such considerations into the initial stages of panel production creates materials that are not only environmentally friendly during their use but also more conducive to efficient recycling at the end of their life cycle.

6 Conclusions

This chapter explores the environmental impact of green lignocellulosic panels, emphasizing their ecological footprint. Understanding the environmental implications involves examining the life cycle, with LCA providing a holistic perspective. Green panels' multifaceted nature, covering material selection, sustainable sourcing, eco-friendly manufacturing, and recycling challenges, is revealed through LCA. Material selection is crucial, with a shift from synthetic to bio-based binders promoting environmental friendliness. Greener pretreatment technologies in manufacturing, like supercritical fluids, reduce waste and energy consumption. Product manufacturing, using technologies like fiber mat and compounding, refines the green composite. Closed molding processes, being more environmentally friendly, offer lower emissions, reduced waste, and higher efficiency. Green lignocellulosic panels, derived from renewable sources, showcase inherent sustainability. However, recycling poses challenges due to adhesives and binders. Researchers seek innovative solutions, influenced by the circular economy concept. Proactive measures, including eco-friendly binders and binderless technology, promote materials conducive to efficient recycling.

Economic perspectives within green panels highlight sustainability, innovation, and resilience. Exploring eco-friendly materials aligns with environmental goals and creates economic opportunities. Efficient raw material procurement and sustainable manufacturing processes contribute to cost-effectiveness and competitiveness. The shift to circular economy principles, supported by legislative policies, addresses environmental challenges, and unlocks economic incentives. In conclusion, this chapter lays the foundation for examining LCA, sustainable raw material sourcing, eco-friendly manufacturing, and recycling challenges. Navigating sustainability complexities sets the stage for informed decision-making and advancements in the quest for truly "green" panels.

References

1. Asyraf, M.R.M., Ishak, M.R., Sapuan, S.M., Yidris, N., Ilyas, R.A., Rafidah, M., Razman, M.R.: Potential application of green composites for cross arm component in transmission tower: a brief review. Int. J. Polym. Sci. **2020**, 8878300 (2020)
2. Baghban, M.H., Mahjoub, R.: Natural kenaf fiber and LC3 binder for sustainable fiber-reinforced cementitious composite: a review. Appl. Sci. **10**, 357 (2020)
3. Bekhta, P., Chernetskyi, O., Kusniak, I., Bekhta, N., Bryn, O.: Selected properties of plywood bonded with low-density polyethylene film from different wood species. Polymers (Basel) **14**, 51 (2022)
4. Besserer, A., Troilo, S., Girods, P., Rogaume, Y., Brosse, N.: Cascading recycling of wood waste: a review. Polymers (Basel) **13**, 1752 (2021)
5. BFG International: Open vs closed moulding (2024). https://www.bfginternational.com/what-we-do/open-vs-closed-moulding. Accessed 1 Jan 2024

6. Carcassi, O.B., Minotti, P., Habert, G., Paoletti, I., Claude, S., Pittau, F.: Carbon footprint assessment of a novel bio-based composite for building insulation. Sustainability **14**, 1384 (2022)
7. Casas-Ledón, Y., Daza Salgado, K., Cea, J., Arteaga-Pérez, L.E., Fuentealba, C.: Life cycle assessment of innovative insulation panels based on eucalyptus bark fibers. J. Clean. Prod. **249**, 119356 (2020)
8. Chotikhun, A., Hiziroglu, S.: Some properties of composite panels manufactured from eastern redcedar (Juniperus virginiana L.) using modified starch as a green binder. J. Nat. Fibers **14**, 541–550 (2017)
9. Dastan, T., Safian, A., Sheikhzadeh, M.: The crashworthiness performance of integrally woven sandwich composite panels made using natural and glass fibers. J. Compos. Mater. **55**, 1279–1292 (2021)
10. Del Menezzi, C., Amirou, S., Pizzi, A., Xi, X., Delmotte, L.: Reactions with wood carbohydrates and lignin of citric acid as a bond promoter of wood veneer panels. Polymers (Basel) **10**, 833 (2018)
11. Ead, A.S., Appel, R., Alex, N., Ayranci, C., Carey, J.P.: Life cycle analysis for green composites: a review of literature including considerations for local and global agricultural use. J. Eng. Fiber Fabr. **16** (2021)
12. Ercan, E., Atar, M., Kucuktuvek, M., Keskin, H.: Characterization of formaldehyde emission and combustion properties of peanut (Arachis Hypogaea) husk-based green composite panels for building applications. Drv. Ind. **73**, 139–149 (2022)
13. Espadas-Aldana, G., Guaygua-Amaguaña, P., Vialle, C., Belaud, J.-P., Evon, P., Sablayrolles, C.: Life cycle assessment of olive pomace as a reinforcement in polypropylene and polyethylene biocomposite materials: a new perspective for the valorization of this agricultural by-product. Coatings **11**, 525 (2021)
14. Faruk, O., Bledzki, A.K., Fink, H.-P., Sain, M.: Biocomposites reinforced with natural fibers: 2000–2010. Prog. Polym. Sci. **37**, 1552–1596 (2012)
15. François, C., Placet, V., Beaugrand, J., Pourchet, S., Boni, G., Champion, D., Fontaine, S., Plasseraud, L.: Can supercritical carbon dioxide be suitable for the green pretreatment of plant fibres dedicated to composite applications? J. Mater. Sci. **55**, 4671–4684 (2020)
16. Fu, Q., Zhuang, B., Wang, X.M., Cloutier, A., Roussière, F., Bouffard, J.F.: Thermo-hydrolytic recycling of urea-formaldehyde resin-bonded laminated particleboards. BioResources **17**, 6847–6859 (2022)
17. Hassanin, A.H., Hamouda, T., Candan, Z., Kilic, A., Akbulut, T.: Developing high-performance hybrid green composites. Compos. Part B Eng. **92**, 384–394 (2016)
18. Hauschild, M.Z.: Introduction to LCA methodology. In: Hauschild, M.Z., Rosenbaum, R.K., Olsen, S.I. (eds.) Life Cycle Assessment: Theory and Practice, pp. 59–66. Springer International Publishing, Cham (2018)
19. Huijbregts, M.A.J., Steinmann, Z.J.N., Elshout, P.M.F., Stam, G., Verones, F., Vieira, M., Zijp, M., Hollander, A., van Zelm, R.: ReCiPe2016: a harmonised life cycle impact assessment method at midpoint and endpoint level. Int. J. Life Cycle Assess. **22**, 138–147 (2017)
20. International Organization for Standardization: Environmental management—life cycle assessment—principles and framework. Geneva (2006)
21. International Organization for Standardization: ISO. Environmental management—Life cycle assessment—Requirements and guidelines. Geneva (2006)
22. Kc, B., Pervaiz, M., Faruk, O., Tjong, J., Sain, M.: Green composite manufacturing via compression molding and thermoforming. In: Salit, M.S., Jawaid, M., Bin, Y.N., Hoque, M.E. (eds.) Manufacturing of Natural Fibre Reinforced Polymer Composites, pp. 45–63. Springer International Publishing, Cham (2015)
23. Kharazipour, A., Kües, U.: Recycling of wood composites and solid wood products. In: Kües, U. (ed.) Wood Production, Wood Technology, and Biotechnological Impacts, pp. 509–533. Universitätsverlag Göttingen (2007)
24. Lee, S.H., Md Tahir, P., Lum, W.C., Tan, L.P., Bawon, P., Park, B.-D., Osman Al Edrus, S.S., Abdullah, U.H.: A review on citric acid as green modifying agent and binder for wood. Polymers (Basel) **12**, 1692 (2020)

25. Mahmood, H., Moniruzzaman, M., Yusup, S., Akil, H.M.: Pretreatment of oil palm biomass with ionic liquids: a new approach for fabrication of green composite board. J. Clean. Prod. **126**, 677–685 (2016)
26. Mahmood, H., Moniruzzaman, M., Yusup, S., Welton, T.: Ionic liquids assisted processing of renewable resources for the fabrication of biodegradable composite materials. Green Chem. **19**, 2051–2075 (2017)
27. Mbow, C., Smith, P., Skole, D., Duguma, L., Bustamante, M.: Achieving mitigation and adaptation to climate change through sustainable agroforestry practices in Africa. Curr. Opin. Environ. Sustain. **6**, 8–14 (2014)
28. Mewoli, A.E., Segovia, C., Njom, A.E., Ebanda, F.B., Eyinga Biwôlé, J.J., Xinyi, C., Ateba, A., Girods, P., Pizzi, A., Brosse, N.: Characterization of tannin extracted from Aningeria altissima bark and formulation of bioresins for the manufacture of Triumfetta cordifolia needle-punched nonwovens fiberboards: novel green composite panels for sustainability. Ind. Crops Prod. **206**, 117734 (2023)
29. Moll, L., Wever, C., Völkering, G., Pude, R.: Increase of miscanthus cultivation with new roles in materials production—a review. Agronomy **10**, 308 (2020)
30. Muthu, S.S.: 6 - Estimating the overall environmental impact of textile processing. In: Muthu SSBT-A the EI of T and the CSC (Second E) (eds.) Life Cycle Assessment of Textile Products. The Textile Institute Book Series, pp. 105–129. Woodhead Publishing (2020)
31. Oliveira, P.R., May, M., Panzera, T.H., Hiermaier, S.: Bio-based/green sandwich structures: a review. Thin-Walled Struct. **177**, 109426 (2022)
32. Opacic, L., Sowlati, T.: Applications of discrete-event simulation in the forest products sector: a review. For. Prod. J. **67**, 219–229 (2017)
33. Operato, L., Vitiello, L., Aprea, P., Ambrogi, V., Salzano de Luna, M., Filippone, G.: Life cycle assessment of poly(lactic acid)-based green composites filled with pine needles or kenaf fibers. J. Clean. Prod. **387**, 135901 (2023)
34. Owodunni, A.A., Lamaming, J., Hashim, R., Abdulwahab Taiwo, O.F., Hussin, M.H., Mohamad Kassim, M.H., Bustami, Y., Sulaiman, O., Mohamad Amini, M.H., Hiziroglu, S.: Properties of green particleboard manufactured from coconut fiber using a potato starch based adhesive. BioResources **15**, 2279–2292 (2020)
35. Pan, Y., Birdsey, R.A., Fang, J., Houghton, R., Kauppi, P.E., Kurz, W.A., Phillips, O.L., Shvidenko, A., Lewis, S.L., Canadell, J.G., Ciais, P., Jackson, R.B., Pacala, S.W., McGuire, A.D., Piao, S., Rautiainen, A., Sitch, S., Hayes, D.: A large and persistent carbon sink in the world's forests. Science **(80-) 333**, 988–993 (2011)
36. Pang, B., Zhou, T., Cao, X.-F., Zhao, B.-C., Sun, Z., Liu, X., Chen, Y.-Y., Yuan, T.-Q.: Performance and environmental implication assessments of green bio-composite from rice straw and bamboo. J. Clean. Prod. **375**, 134037 (2022)
37. Rainforest Alliance: What is sustainable forestry? Insights (2016). https://www.rainforest-alliance.org/insights/what-is-sustainable-forestry/. Accessed 22 Nov 2023
38. Roy, R., Rahman, M.S., Raynie, D.E.: Recent advances of greener pretreatment technologies of lignocellulose. Curr. Res. Green Sustain. Chem. **3**, 100035 (2020)
39. Sanford, S.S., Finley, J., Jacobson, M.: Sustainable forestry. From woods, pp. 1–4 (2002)
40. Secretariat of the Convention on Biological Diversity: Sustainable forest management, biodiversity, and livelihoods: a good practice guide. Montreal (2009)
41. Sommerhuber, P.F., Wenker, J.L., Rüter, S., Krause, A.: Life cycle assessment of wood-plastic composites: analysing alternative materials and identifying an environmental sound end-of-life option. Resour. Conserv. Recycl. **117**, 235–248 (2017)
42. Vitrone, F., Ramos, D., Ferrando, F., Salvadó, J.: Binderless fiberboards for sustainable construction. Materials, production methods and applications. J Build Eng **44**, 102625 (2021)
43. Yan, L., Ma, R., Li, L., Fu, J.: Hot water pretreatment of lignocellulosic biomass: an effective and environmentally friendly approach to enhance biofuel production. Chem. Eng. Technol. **39**, 1759–1770 (2016)
44. Yuan, Y., Guo, M.: Do green wooden composites using lignin-based binder have environmentally benign alternatives? A preliminary LCA case study in China. Int. J. Life Cycle Assess. **22**, 1318–1326 (2017)

45. Zhao, X., Copenhaver, K., Wang, L., Korey, M., Gardner, D.J., Li, K., Lamm, M.E., Kishore, V., Bhagia, S., Tajvidi, M., Tekinalp, H., Oyedeji, O., Wasti, S., Webb, E., Ragauskas, A.J., Zhu, H., Peter, W.H., Ozcan, S.: Recycling of natural fiber composites: challenges and opportunities. Resour. Conserv. Recycl. **177**, 105962 (2022)
46. Zimmer, A., Angie Lunelli Bachmann, S.: Challenges for recycling medium-density fiberboard (MDF). Results Eng. **19**, 101277 (2023)

MIX
Papier aus verantwortungsvollen Quellen
Paper from responsible sources
FSC® C105338

If you have any concerns about our products,
you can contact us on
ProductSafety@springernature.com

In case Publisher is established outside the EU,
the EU authorized representative is:
Springer Nature Customer Service Center GmbH
Europaplatz 3, 69115 Heidelberg, Germany

Printed by Libri Plureos GmbH
in Hamburg, Germany